Ben Stacy Jerrik (Ed.)

Healaugh

Ben Stacy Jerrik (Ed.)

Healaugh

Swaledale, Civil parishes in England, Yorkshire Dales, Richmondshire

Part Press

Imprint

Publisher:
Part Press is a trademark of
International Book Market Service Ltd., 17 Rue Meldrum, Beau Bassin, 1713-01 Mauritius
Email: info@bookmarketservice.com
Website: www.bookmarketservice.com

Published in 2012

Printed in: U.S.A., U.K., Germany. This book was not produced in Mauritius.

ISBN: 978-613-6-05457-5

Contents

Articles

Healaugh	1
Swaledale	2
Civil_parishes_in_England	4
Yorkshire_Dales	10
Richmondshire	15
North_Yorkshire	19
Reeth	27
Telephone_booth	29
Tadcaster	34
Reeth,_Fremington_and_Healaugh	40

References

Article Sources and Contributors	41
Image Sources, Licenses and Contributors	42

Healaugh

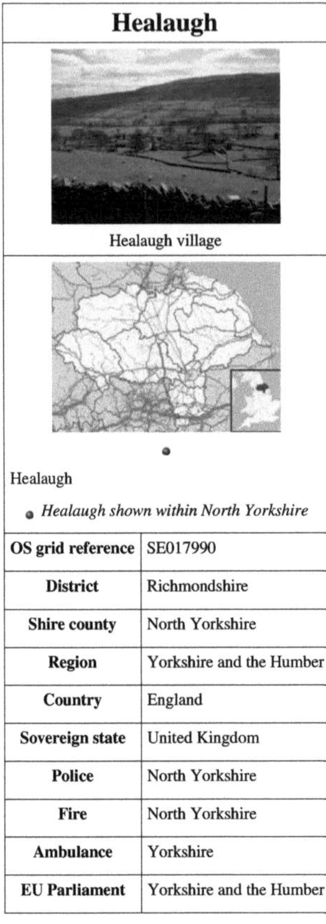

Healaugh	
Healaugh village	
Healaugh Healaugh shown within North Yorkshire	
OS grid reference	SE017990
District	Richmondshire
Shire county	North Yorkshire
Region	Yorkshire and the Humber
Country	England
Sovereign state	United Kingdom
Police	North Yorkshire
Fire	North Yorkshire
Ambulance	Yorkshire
EU Parliament	Yorkshire and the Humber

Healaugh (pronounced "hee-law")[1] is a small village and civil parish in Swaledale in the Yorkshire Dales. It is in the Richmondshire district of North Yorkshire, England and lies about 1 mile west of Reeth.

The name *Healaugh* is derived from a Saxon word (*Heah*) meaning a *high-level forest clearing*.[2]

The village is small, with no amenities except a stone trough fed by a hillside stream,[1] and the village telephone box. The latter is unusually well endowed, with a carpet, waste paper bin, ash tray, directories and fresh flowers. Visitors may leave a donation.[2]

Note

Confusingly, there is another small Yorkshire village called *Healaugh* just north of Tadcaster, in the Selby district of North Yorkshire.

See also

- Reeth, Fremington and Healaugh

References

[1] "Herriot Way" (http://www.herriotway.co.uk/place.php3?place=Healaugh&p=7&walk=herriotway). *Sherpa Expeditions Limited*. 2007. . Retrieved 2007-06-29.

[2] Dalesman Publishing Ltd (2002-04-09). "Healaugh and the River Swale from Reeth" (http://www.dalesman.co.uk/walks/healaugh.htm). *Dalesman magazine*. . Retrieved 2007-06-29.

Swaledale

Swaledale is one of the northernmost dales (valleys) in the Yorkshire Dales National Park in northern England. It is the dale of the River Swale on the east side of the Pennines in North Yorkshire.

Swaledale

Geographical overview

Swaledale starts to the east of Nine Standards Rigg, the prominent ridge with nine ancient tall cairns on the Cumbria–Yorkshire boundary which forms part of the main east–west watershed of Northern England. To the west lies Kirkby Stephen and the Westmorland Limestone Plateau.

The moors on the eastern flank of the Rigg's moorland become more and more concave as they descend, to become the narrow valley sides of upper Swaledale at the small village of Keld. From there, the valley runs briefly south then turns east at Thwaite to broaden progressively as it passes Muker, Gunnerside and Reeth. The Pennine valley ends at the market town of Richmond, where an important medieval castle still watches the important ford from the top of a cliff. Below Richmond, the valley sides flatten out and the Swale flows across lowland farmland to meet the Ure just east of Boroughbridge at a point known as Swale Nab. The Ure becomes the Ouse, and eventually (on merging with the Trent) the Humber.

From the north, Arkengarthdale and its river the Arkle Beck join Swaledale at Reeth. To the south, Wensleydale, home of the famous Wensleydale cheese, runs parallel with Swaledale. The two dales are separated by a ridge including Great Shunner Fell, and joined by the road over Buttertubs Pass.

Physical character

Swaledale at East Applegarth, near Richmond

Swaledale is a typical limestone Yorkshire dale, with its narrow valley-bottom road, green meadows and fellside fields, white sheep and white stone walls on the glacier-formed valley sides, and darker moorland skyline. The upper parts of the dale are particularly striking because of its large old limestone field barns and its profusion of wild flowers.[1] The latter are thanks to the return to the practice of leaving the cutting of grass for hay or silage until wild plants have had a chance to seed. Occasionally visible from the valley bottom road are the slowly-fading fellside scars of the 18th and 19th century lead mining industry. Ruined stone mine buildings remain, taking on the same colours as the landscape into which they are crumbling.

Swaledale is home to many small but beautiful waterfalls, such as Richmond Falls, Kisdon Force and Catrake Force.

Agriculture and industry

Sheep-farming has always been central to economic life in Swaledale, which has lent its name to a breed of round-horned sheep. Traditional Swaledale products are woollens and Swaledale cheese, which was formerly made from ewe's milk. These days it is made from cow's milk. During the 19th century, a major industry in the area was lead mining.

Current human activities

Today, tourism has become important, and Swaledale attracts thousands of visitors a year. It is very popular with walkers, particularly because the Coast to Coast Walk passes along it. Unlike Wensleydale it has no large settlements on the scale of Hawes or Leyburn, nor an obvious tourist hook such as former's connection with James Herriot, and so, like Coverdale, it enjoys a quieter tone, especially as it is more remote compared to, say, Wharfedale, which is much further south and easily accessible from the West Yorkshire metropolis.

In May and June every year, Swaledale hosts the two-week long Swaledale Festival, which combines a celebration of small-scale music and a programme of guided walks.[2]

Since 1950, Swaledale has been the host of the Scott Trial, a British motorcycle trials competition run over an off road course of approximately 70 miles, raising money for the "Scott charities", a range of local non-profit making organisations.[3]

See also

- Swaledale Festival
- Swaledale Museum
- Swaledale and Arkengarthdale Archaeology Group [4]

References

[1] (http://www.yorkshirenet.co.uk/yorkshire-dales/swaledale.htm)
[2] "Swaledale Festival" (http://www.swaledale-festival.org.uk/). Swaledale Festival. . Retrieved 2010-03-24.
[3] "Scott Trial History" (http://www.richmondmotorclub.com/scothistory.php). . Retrieved 2009-05-07.
[4] http://www.swaag.org/

Civil_parishes_in_England

Civil parish (England)	
Category	Parish
Location	England
Found in	Districts
Created by	Various, see text
Created	Various, see text
Number	~4,500 (as of 2009)
Possible types	City
	Community
	Neighbourhood
	Parish
	Town
	Village
Government	City council
	Community council
	Neighbourhood council
	Parish council
	Town council
	Village council

In England, a **civil parish** is a territorial designation and, where they are found, the lowest tier of local government below districts and counties. It is an administrative parish, in comparison to an ecclesiastical parish.

A civil parish can alternatively be known as a **town**, **village**, **neighbourhood** or **community** by resolution of its parish council; and in a limited number of cases has city status granted by the monarch. They cover only part of England, corresponding to 35% of the population.

There are currently no civil parishes in Greater London and before 2008 their creation was not permitted within a London borough.[1]

History

Ancient origins

The division into ancient parishes was linked to the manorial system, with parishes and manors often sharing the same boundaries.[2] Initially the manor was the principal unit of local administration and justice in the early rural economy. Eventually the church replaced the manor court as the rural administrative centre and levied a local tax on produce known as a tithe.[2] Responsibilities such as relief of the poor passed from the Lord of the Manor to the church, although in practice it was administered by monasteries. Following the dissolution of the monasteries, the power to levy a rate to fund relief of the poor was conferred on the parish authorities by the 1601 Act for the Relief of the Poor.

The parish authorities were known as vestries and consisted of all the inhabitants of the parish. As the population was growing it became increasingly difficult to convene meetings as an open vestry. In some, mostly built up, areas the select vestry took over responsibility from the community at large. This innovation improved efficiency, but allowed governance by a self-perpetuating elite.[2] The administration of the parish system relied on the monopoly of the English church. As religious membership became more fractured, such as through the revival of Methodism, the legitimacy of the parish vestry came into question and the perceived inefficiency and corruption inherent in the system became a source for concern.[2] Because of this scepticism, during early the 19th century the parish progressively lost its powers to ad-hoc boards and other organisations, such as the loss of responsibility for poor relief through the Poor Law Amendment Act 1834. The replacement boards were each able to levy their own rate in the parish. The church rate ceased to be levied in many areas and was abolished altogether in 1868.[2]

Civil and ecclesiastical split

The ancient parishes diverged into two distinct units during the 19th century. The Poor Law Amendment Act 1866 declared all areas that levied a separate rate —extra-parochial areas, townships, and chapelries— become civil parishes as well. The parishes for church use continued unchanged as ecclesiastical parishes. The latter part of the 19th century saw most of the ancient irregularities inherited by the civil system cleaned up, with the majority of exclaves abolished.

Reform

Civil parishes in their modern sense were established afresh in 1894, by the Local Government Act 1894. The Act abolished vestries, and established elected parish councils in all rural civil parishes with more than 300 electors. These were grouped into rural districts. Boundaries were altered to avoid parishes being split between counties. Urban parishes continued to exist, and were generally coterminous with the urban district or municipal borough in which they were situated. Large towns originally split between multiple parishes were, for the most part, eventually consolidated into one parish. No parish councils were formed for urban parishes, and their only function was as areas electing guardians to Poor Law Unions. With the abolition of the poor law system in 1930 the parishes had only a nominal existence.

In 1965 civil parishes in London were formally abolished when Greater London was created, as the legislative framework for Greater London did not make provision for any local government body below a London borough (since all of London was previously part of a metropolitan borough, municipal borough or urban district, no actual parish councils were abolished). In 1974 the Local Government Act 1972 retained civil parishes in rural areas and small urban areas, but abolished them in larger urban areas. Many former urban districts and municipal boroughs that were being abolished rather than succeeded were continued as new parishes. Urban areas that were considered too large to be single parishes were refused this permission and became unparished areas. The Act also led to the possibility of sub-division of all districts (apart from London boroughs, reformed in 1965), into multiple civil parishes. For example, Oxford, whilst entirely unparished in 1974, now has four civil parishes, covering part of its area.

Revival

The creation of town and parish councils is encouraged in unparished areas. The Local Government and Rating Act 1997 created a procedure which gave local residents the right to demand that a new parish and council be created in unparished areas.[3] This was extended to London boroughs by the Local Government and Public Involvement in Health Act 2007[4] - with this, the City of London is at present the only part of England where civil parishes cannot be created.

If a sufficient number of electors in an area of a proposed new parish (ranging from 50% in an area with less than 500 electors to 10% in one with more than 2,500) sign a petition demanding its creation, then the local district council or unitary authority must consider the proposal.[1] Recently established parish councils include Daventry (2003), Folkestone (2004), and Brixham (2007). In 2003 seven new parish councils were set up for Burton upon Trent, and in 2001 the Milton Keynes urban area became entirely parished, with ten new parishes being created. In 2003, the village of Great Coates (Grimsby) regained parish status. Parishes can also be abolished where there is evidence that this in response to "justified, clear and sustained local support" from the area's inhabitants.[1] Examples include Birtley, which was abolished in 2006 and Southsea abolished in 2010.[5] [6]

Governance

Every civil parish has a parish meeting, consisting of all the electors of the parish. Generally a meeting is held once a year. A civil parish may have a parish council which exercises various local responsibilities given by statute. If a parish has fewer than 200 electors it is usually deemed too small to have a parish council, and instead will only have a parish meeting; an example of direct democracy. Alternatively several small parishes can be grouped together and share a common parish council, or even a common parish meeting. In places where there is no civil parish (unparished areas), the administration of the activities normally undertaken by the parish becomes the responsibility of the district or borough council. According to the Government's Department for Communities and Local Government, in England in 2011 there are 9,946 parishes. [7] Since 1997 around 100 new civil parishes have been created, in some cases splitting existing civil parishes, but mostly by creating new ones from unparished areas.

Powers and functions

Typical activities undertaken by parish or town councils include:[8]

- The provision and upkeep of certain local facilities such as allotments, bus shelters, parks, playgrounds, public seats, public toilets, public clocks, street lights, village or town halls, and various leisure and recreation facilities.
- Maintenance of footpaths, cemeteries and village greens
- Since 1997 parish councils have had new powers to provide community transport (such as a minibus), crime prevention measures (such as CCTV) and to contribute money towards traffic calming schemes.
- Parish councils are supposed to act as a channel of local opinion to larger local government bodies, and as such have the right to be consulted on any planning decisions affecting the parish.
- Giving of grants to local voluntary organisations, and sponsoring public events, including entering Britain in Bloom.

The role played by parish councils varies. Smaller parish councils have only limited resources and generally play only a minor role, while some larger parish councils have a role similar to that of a small district council. Parish councils receive funding by levying a "precept" on the council tax paid by the residents of the parish.

Councillors and elections

Parish councils are run by volunteer councillors who are elected to serve for four years and are not paid. Some councils have chosen to pay their elected members a small allowance as permitted under Part 5 of the Local Government Act 2000 The Local Authorities (Members' Allowances) (England) Regulations 2003.[9] The number of councillors varies roughly in proportion to the population of the parish. Most parish councillors are elected to represent the entire parish, though in parishes with larger populations or those that cover large areas, the parish can be divided into wards. These wards then return a certain number of councillors each to the parish council (depending on their population). Only if there are more candidates standing for election than there are seats on the council will an election be held. However, sometimes there are fewer candidates than seats. When this happens, the vacant seats have to be filled by co-option by the council. When a vacancy arises for a seat mid-term, an election is only held if a certain number (usually 10) of parish residents request an election. Otherwise the council will co-opt someone to be the replacement councillor. Every Parish Council in England must adopt a code of conduct, and parish councillors must comply with its standards, enforced by the Standards Board for England.

Status and styles

A parish can gain city status but only if that is granted by the Crown. In England, there are currently eight parishes with city status, all places with long-established Anglican cathedrals: Chichester, Ely, Hereford, Lichfield, Ripon, Salisbury, Truro and Wells.

The council of an ungrouped parish may unilaterally pass a resolution giving the parish the status of a town.[10] The parish council becomes a "town council".[11] Around 400 parish councils are called town councils.

Under the Local Government and Public Involvement in Health Act 2007, a civil parish may now be given an "alternative style" meaning one of the following:

- community
- neighbourhood
- village

The chairman of a town council will have the title "town mayor" and that of a parish council which is a city will usually have the title of mayor. As a result, a parish council can also be called a town council, a community council, a village council or occasionally a city council (though most cities are not parishes but principal areas, or in England specifically metropolitan boroughs, non-metropolitan districts).[12] [13]

Charter trustees

When a city or town has been abolished as a borough, and it is considered desirable to maintain continuity of the charter, the charter may be transferred to a parish council for its area. Where there is no such parish council, the district council may appoint charter trustees to whom the charter and the arms of the former borough will belong. The charter trustees (who consist of the councillor or councillors for the area of the former borough) maintain traditions such as mayoralty. An example of such a city was Hereford, whose city council was merged in 1998 to form a unitary Herefordshire. The area of the city of Hereford remained unparished until 2000 when a parish council was created for the city. The charter trustees for the City of Bath make up the majority of the councillors on Bath and North East Somerset Council.

Geography

Civil parishes do not cover the whole of England, with none in Greater London and very few in the other conurbations. Civil parishes vary greatly in size: many cover tiny hamlets with populations of less than 100, whereas some large parishes cover towns with populations of tens of thousands. Weston-super-Mare, with a population of 71,758, is the most populous civil parish. In many cases, several small villages are located in a single parish. Large urban areas are mostly unparished, as the government at the time of the Local Government Act 1972 discouraged their creation for large towns or their suburbs, but there is generally nothing to stop their establishment. For example, Birmingham has just one parish, New Frankley, whilst Oxford has four, and Northampton has seven. Parishes could not however be established in London until the changing of the law in 2007 and as yet none have been established there.

Deserted parishes

The 2001 census recorded several parishes with no inhabitants. These were Chester Castle (in the middle of Chester city centre), Newland with Woodhouse Moor, Beaumont Chase, Martinsthorpe, Meering, Stanground North (subsequently abolished), Sturston, Tottington, and Tyneham. The last three had been taken over by the British Armed Forces during World War II and remain deserted.

Detached parts and divided parishes

Ancient parishes often had detached parts, exclaves and enclaves which were not contiguous with the rest of the parish. In some cases the detached part was in a different county. In other cases, an entire parish was in a detached part of the county to which it belonged. There were also many examples of parishes divided between two or more counties.

These anomalies were mostly addressed in the 19th century. Before civil parishes were introduced, the Counties (Detached Parts) Act 1844 transferred many (but not all) parishes which were detached parts of a county to the county in which they were geographically located. The remaining detached parishes were transferred in the 1890s and in 1931. The detached part of the parish of Tetworth, Huntingdonshire, surrounded by Cambridgeshire, remained until the boundaries were changed in 1965.

Other legislation, including the Divided Parishes and Poor Law Amendment Act 1882, eliminated most instances of civil parishes belonging to two (or more) counties, and by 1901 Stanground in Huntingdonshire and the Isle of Ely was the sole remaining example.[14] Stanground was split into two parishes, one in each county, in 1905.[15]

See also

- List of civil parishes in England

References

- Wright, R S; Hobhouse, Henry (1884). *An Outline of Local Government and Local Taxation in England and Wales (Excluding the Metropolis)*. London: W Maxwell & Son.

[1] *Guidance on Community Governance Reviews* (http://www.communities.gov.uk/documents/localgovernment/pdf/1527635.pdf). London: Department for Communities and Local Government. 2010. ISBN 9781409824213. .

[2] Arnold-Baker, Charles (1989). *Local Council Administration in English Parishes and Welsh Communities*. Longcross Press. ISBN 9780902378094.

[3] *What is a parish or town council*, National Association of Local Councils website, accessed 14 August 2010 (http://www.nalc.gov.uk/About_NALC/What_is_a_parish_or_town_council/All_about_local_councils.aspx)

[4] This was ss.58-77 of bill, which received royal assent on 30 October 2007: http://www.publications.parliament.uk/pa/pabills/200607/local_government_and_public_involvement_in_health.htm

[5] "Birtley Town Council – Annual Return 2005/2006". Gateshead Council. 29 September 2006.

[6] "The Portsmouth City Council (Reorganisation of Community Governance) Order 2010" (http://www2.portsmouth.co.uk/pdfs/publicnotices/2010/160410.pdf). . Retrieved 11 September 2010.

[7] Parishes and Charter Trustees in England 2011-12 (http://www.communities.gov.uk/publications/corporate/statistics/parishes201112)

[8] Full list of powers of parish councils (http://www.nalc.gov.uk/Document/Download.aspx?uid=079806d3-0e69-4588-9a32-ef02c91598df) - Downloadable Microsoft Word Document

[9] Local Government Act 2000 The Local Authorities (Members' Allowances) (England) Regulations 2003 Reg 30 (http://www.statutelaw.gov.uk/content.aspx?LegType=All+Primary&PageNumber=1&NavFrom=2&parentActiveTextDocId=0&linkToSearchEnacted=0&linkToSearchDay=3&linkToLocale=E+W+S+N.I.&linkToMatchExactLocale=0&linkToSearchMonth=12&linkToSearchYear=2008&linkToATDocumentId=840810&filesize=133430)

[10] "The council of a parish which is not grouped with any other parish may resolve that the parish shall have the status of a town" "Local Government Act 1972 (c.70), Part XIII" (http://www.opsi.gov.uk/RevisedStatutes/Acts/ukpga/1972/cukpga_19720070_en_35#pt13-pb1-l1g255). *Revised Statutes*. Office of Public Sector Information. 1972. . Retrieved 11 September 2010.

[11] *Local government in England and Wales: A Guide to the New System*. London: HMSO. 1974. p. 158. ISBN 0117508470.

[12] NALC - National Association of Local Councils (http://www.nalc.gov.uk/About_NALC/What_is_a_parish_or_town_council/What_is_a_council.aspx) Retrieved 26 December 2009

[13] Guidance on Community Governance Reviews (April 2008), Local Government publication ISBN 978-8511-2917-1 (http://www.communities.gov.uk/publications/localgovernment/communitygovernancereviews) Retrieved 26 December 2009

[14] 1901 Census of England and Wales, General Report: Administrative Counties and County Boroughs (http://www.visionofbritain.org.uk/text/chap_page.jsp?t_id=SRC_P&c_id=3&cpub_id=EW1901GEN)

[15] Local Government Board Order No. 56410, made under the Local Government Act 1894 (56 & 57 Vict. c.73) s.36

External links

- In praise of ... civil parishes (http://www.guardian.co.uk/commentisfree/2011/may/16/in-praise-of-civil-parishes) Editorial in *The Guardian*, 2011-05-16.

Yorkshire_Dales

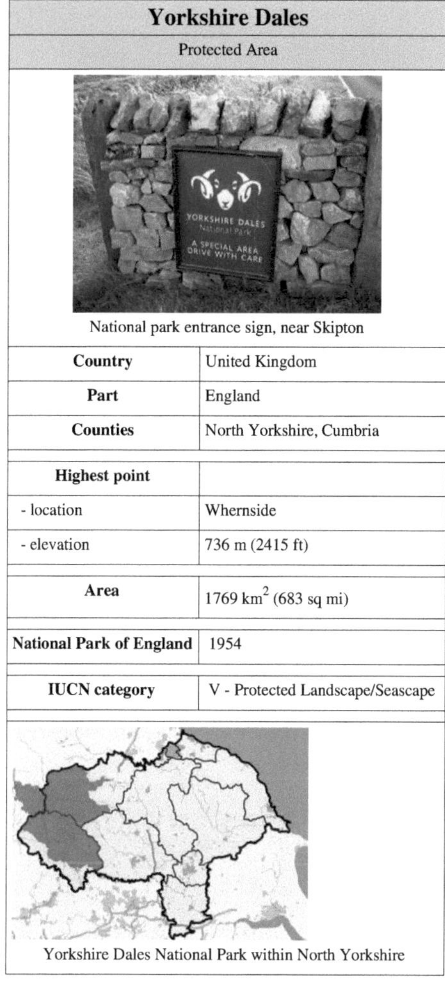

Yorkshire Dales	
Protected Area	
National park entrance sign, near Skipton	
Country	United Kingdom
Part	England
Counties	North Yorkshire, Cumbria
Highest point	
- location	Whernside
- elevation	736 m (2415 ft)
Area	1769 km^2 (683 sq mi)
National Park of England	1954
IUCN category	V - Protected Landscape/Seascape
Yorkshire Dales National Park within North Yorkshire	

The **Yorkshire Dales** is the name given to an upland area in Northern England.

The area lies within the historic county boundaries of Yorkshire, though it spans the ceremonial counties of North Yorkshire, West Yorkshire and Cumbria. Most of the area falls within the Yorkshire Dales National Park, created in 1954, and now one of the fifteen National parks of Britain, but the term also includes areas to the east of the National Park, notably Nidderdale.

The Dales is a collection of river valleys and the hills among them, rising from the Vale of York westwards to the hilltops of the main Pennine watershed (the British English meaning). In some places the area even extends westwards across the watershed, but most of the valleys drain eastwards to the Vale of York, into the Ouse and then the Humber.

The word *dale* comes from the Nordic/Germanic word for valley (*dal*, *tal*), and occurs in valley names across Yorkshire (and Northern England generally) but the name Yorkshire Dales is generally used to refer specifically to the dales west of the Vale of York and north of the West Yorkshire Urban Area. The Yorkshire Dales is served by its own radio station, Fresh Radio, which broadcasts programmes from studio bases in Skipton and Richmond.

Geography

Most of the dales in the Yorkshire Dales are named after their river or stream (e.g. Arkengarthdale, formed by Arkle Beck). The best-known exception to this rule is Wensleydale, which is named after the small village (but former market town) of Wensley rather than the River Ure, although an older name for the dale is Yoredale. In fact, valleys all over Yorkshire are called "(name of river)+dale"—but only the more northern Yorkshire valleys (and only the upper, rural, reaches) are included in the term "The Dales". For example, the southern boundary area lies in Wharfedale and Airedale. The lower reaches of these valleys are not usually included in the area and Calderdale much further south, would not normally be referred to as part of "The Dales" even though it is a dale, is in Yorkshire, and the upper reaches are as scenic and rural as many valleys further north.

Geographically, the classical Yorkshire Dales spread to the north from the market and spa towns of Settle, Skipton, Ilkley and Harrogate in North Yorkshire, with most of the larger southern dales (e.g. Ribblesdale, Malhamdale and Airedale, Wharfedale and Nidderdale) running roughly parallel from north to south. The more northerly dales (e.g. Wensleydale, Swaledale and Teesdale) run generally from west to east. There are also many other smaller or lesser known dales such as Arkengarthdale, Bishopdale, Clapdale, Coverdale, Kingsdale, Littondale, Langstrothdale, Raydale, Waldendale and the Washburn Valley whose tributary streams and rivers feed into the larger valleys, and Barbondale, Dentdale, Deepdale and Garsdale which feed west to the River Lune.

The characteristic scenery of the Dales is green upland pastures separated by dry-stone walls and grazed by sheep and cattle. The dales themselves are 'U' and 'V' shaped valleys, which were enlarged and shaped by glaciers, mainly in the most recent, Devensian ice age. The underlying rock is principally Carboniferous Limestone (which results in a number of areas of limestone pavement) in places interspersed with shale and sandstone and topped with Millstone Grit. To the north and west of the Dent Fault, the hills are principally formed from older Silurian and Ordovician rocks, which make up the Howgill Fells.

Many of the upland areas consist of heather moorland, used for grouse shooting in the months following 12 August each year (the 'Glorious Twelfth').

Cliffs of Carboniferous Limestone are a common geological feature in the Yorkshire Dales, this panoramic image shows the western face of Thwaites Scars taken from Long Lane.

Tourism

The majority of visitors come to sightsee, with 75% visiting to drive around and 65% walking around. This indicates that most people visiting are there to take in the beauty of the surroundings. 26% also partake in nature trails and spotting wildlife. 45% visit an information centre while 35% visit a castle or other historic site. 94% of visitors travel in a private mode of transport, with 90% using a car. The remaining 6% travelled using public transport.[1]

Cave systems

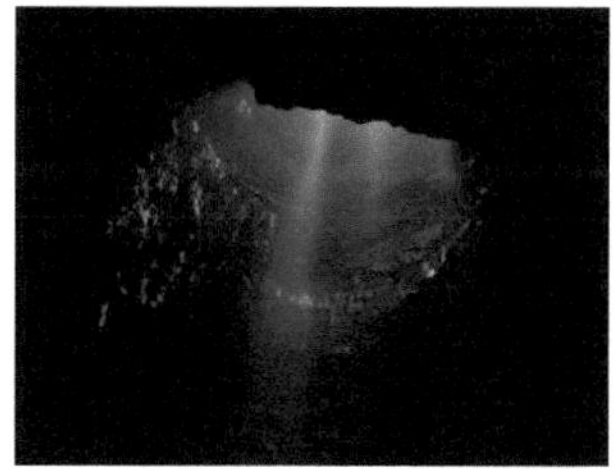
Gaping Gill

Because of the limestone that runs throughout the Dales, there are extensive cave systems present across the area, making it one of the major areas for caving in the UK. Some of these are open to the public for cave tours.[2]

The systems include:

- Gaping Gill System
- Alum Pot System
- Mossdale Caverns
- Kingsdale Caverns
- Leck Fell Caves
- Easegill System
- White Scar Caves in Chapel-le-Dale near Ingleton,[3]
- Ingleborough Cave[4] in Clapdale near Clapham
- Stump Cross Caverns[5] near Pateley Bridge.

Yorkshire Dales National Park

Stone houses in Hawes, a typical example of Dales architecture

In 1954 an area of 1770 square kilometres (**unknown operator: u'strong'** sq mi) was designated the **Yorkshire Dales National Park**. Most of the National Park is in North Yorkshire, though part lies within Cumbria. However, the whole park lies within the historic boundaries of Yorkshire, divided between the North Riding and the West Riding. The park is 50 miles (**unknown operator: u'strong'** km) north east of Manchester; Leeds and Bradford lie to the south, while Kendal is to the west, Darlington to the northeast and Harrogate to the southeast.[6] A proposed westward extension to the park would encompass much of the area between the current park and the M6 motorway, coming close to the towns of Kirkby Lonsdale, Kirkby Stephen and Appleby-in-Westmorland.[7] This proposal to add 162 square miles to the park has now been agreed by all interested parties and merely awaits ministerial approval. For the first time the Yorkshire Dales NP and the Lake District NP will be contiguous.[8]

Over 20,000 residents live and work in the park, which attracts over eight million visitors every year.[9] The area has a large collection of activities for visitors. For example, many people come to the Dales for walking or exercise. The National Park is crossed by several long-distance routes including the Pennine Way, the Dales Way, the Coast to Coast Path and the latest national trail — the Pennine Bridleway.[10] Cycling is also popular and there are several cycleways.[11]

Limestone hills and dry-stone walls in the west of the Yorkshire Dales. This part of the national park is popular with walkers due to the presence of the Yorkshire three peaks.

The Park has its own museum, the Dales Countryside Museum, housed in a conversion of the Hawes railway station in Wensleydale in the north of the area.[12] The park has five visitor centres located in major destinations in the park.[13] These are at:

- Aysgarth Falls
- Grassington
- Hawes
- Malham
- Reeth

Other places and sights within the National Park include:

- Bolton Castle
- Clapham
- Cautley Spout waterfall
- Gaping Gill
- Hardraw Force
- Horton in Ribblesdale
- Kisdon Force (waterfall) in Swaledale
- Malham Cove and Gordale Scar
- Sedbergh
- Settle
- Settle and Carlisle Railway including the Ribblehead Viaduct
- The Yorkshire three peaks

List of Dales

Janet's Foss, near Malham

- Airedale
- Arkengarthdale
- Barbondale
- Birkdale
- Bishopdale
- Chapel-le-Dale
- Coverdale
- Dentdale
- Garsdale
- Langstrothdale
- Littondale
- Kingsdale
- Malhamdale
- Nidderdale
- Ribblesdale
- Swaledale
- Wensleydale

- Wharfedale

See also

- List of peaks in the Yorkshire Dales
- Geology of Yorkshire
- Pennines

The whole of Ingleborough as seen from the peat bog below

References

[1] Yorkshire Dales Tourism Information educational files (http://www.yorkshiredales.org.uk/educationfile08-tourism.pdf)

[2] (http://www.yorkshire.com/outdoors/rocksports/caving-and-show-caves) Caves and Caving in the Yorkshire Dales]

[3] White Scar Caves (http://www.whitescarcave.co.uk/frame.htm)

[4] Ingleborough Cave (http://ingleboroughcave.co.uk/)

[5] Stump Cross Caverns (http://www.stumpcrosscaverns.co.uk/)

[6] Welcome to the Yorkshire Dales National Park (http://www.yorkshiredales.org.uk/)

[7] Natural England - Lakes to Dales Landscape Designation Project (http://www.naturalengland.org.uk/ourwork/conservation/designatedareas/new/northwestdesignationproject/default.aspx)

[8] "BBC News - Yorkshire Dales National Park expansion plans agreed" (http://www.bbc.co.uk/news/uk-england-15082564). 28 September 2011. . Retrieved 29 September 2011.

[9] "Yorkshire Dales National Park Authority - Tourism Education file" (http://www.yorkshiredales.org.uk/educationfile08-tourism.pdf). www.yorkshiredales.org.uk. . Retrieved 2011-09-20.

[10] Yorkshire Dales National Park Authority - Things to do (http://www.yorkshiredales.org.uk/index/enjoying/things_to_do_in_the_dales.htm)

[11] Cycle the Dales (http://www.cyclethedales.org.uk/)

[12] Yorkshire Dales National Park Authority - Dales Countryside Museum (http://www.yorkshiredales.org.uk/index/enjoying/dales_countryside_museum.htm)

[13] Yorkshire Dales National Park Authority - National Park Centres (http://www.yorkshiredales.org.uk/index/enjoying/national_park_centres_1.htm)

External links

- Yorkshire Dales Tourist Board (http://www.yorkshire.com/yorkshire-dales)
- Yorkshire Dales National Park Authority (http://www.yorkshiredales.org.uk/)
- Yorkshire Dales Society (http://www.yds.org.uk)
- Yorkshire Dales Rivers Trust (http://www.yorkshiredalesriverstrust.org.uk/)

Richmondshire

Richmondshire	
— District —	
 coat of arms	
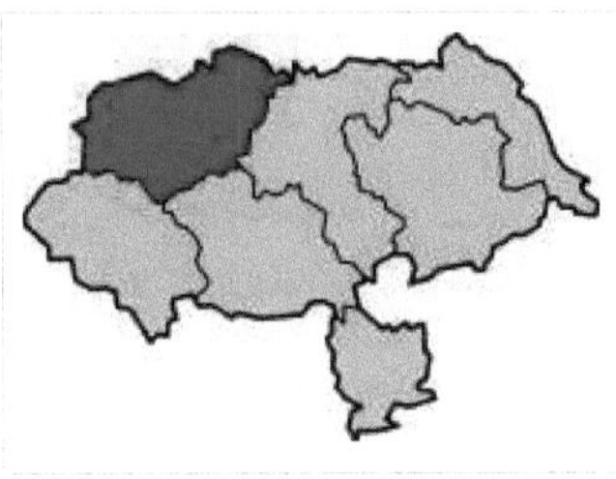 Shown within North Yorkshire	
Sovereign state	United Kingdom
Constituent country	England
Region	Yorkshire and the Humber
Administrative county	North Yorkshire
Founded	
Admin. HQ	Richmond
Government	
• **Type**	Richmondshire District Council
• **Leadership:**	Alternative - Sec.31
• **Executive:**	Independent / Liberal Democrat
• **MPs:**	William Hague
Area	
• **Total**	**unknown operator: u'strong'** sq mi (1319 km^2)
Area rank	13th
Population (2010 est.)	
• **Total**	53,000
• **Rank**	Ranked 318th
Time zone	Greenwich Mean Time (UTC+0)
• **Summer (DST)**	British Summer Time (UTC+1)
Postcode	
ISO 3166-2	

ONS code	36UE
OS grid reference	
NUTS 3	
Ethnicity	97.0% White 1.5% S.Asian[1]
Website	richmondshire.gov.uk [2]

Richmondshire is a local government district of North Yorkshire, England. It covers a large northern area of the Yorkshire Dales including Swaledale and Arkengarthdale, Wensleydale and Coverdale, with the prominent Scots' Dyke and Scotch Corner along the centre. Teesdale lies to the north. It is larger than four of the English counties, such as Berkshire.[3]

History

The history of this district in antiquity is not well known, but the closest important Roman settlement was at Catterick in what became known as Rheged, site of the Battle of Catterick.[4] At the terminus of Scandinavian York, there was a local bout of rebellion in Stainmore, which resulted in the death of Eric Bloodaxe. The Scandinavian settlement of this area was eastwards from the Irish Sea with names such as Gilpatrick in Middleham and Thorfinn in Bedale occurring at the time of the Domesday Book. At the time of the Norman Conquest it was the Fee of Gillingshire, held by Edwin, Earl of Mercia.[5] **Gillingshire**[6] was made up of the Borough of Richmond and five wapentakes of Gilling West, Gilling East, Hang West, Hang East and Hallikeld.[7] After the Harrying of the North, the land became capital of the Duchy of Brittany's Honour of Richmond (first as a barony, then a county and later a dukedom). The honour of Richmond was one of the three largest lordships created by William the Conqueror. He granted it to his cousin, Alan the Red, brother of the Duke of Brittany. Alan had other English estates and by the time that Domesday Book was compiled he was one of the richest and most powerful barons. He died in 1093 and was succeeded by two of his brothers in turn. The family held on to this estate until 1399. Work on the castle started in 1071 after the northern rebellions had died down.

The honour of Richmond comprised 440 manors throughout England. The Yorkshire portion was a compact unit of 199 manors and 43 outlying properties situated near the main roads from Scotland into the Vale of York.[8] Northern England is said to differ from the other areas of the country and the difference between Breton and Norman lordship is seen as being a cause. Richmondshire became an appanage of the English Royal Family during the reign of Edward III of England. In 1525 Henry FitzRoy, 1st Duke of Richmond and Somerset (1519-1536) became Lord Warden of the Marches and Lord President of the Council of the North while living at Sheriff Hutton.

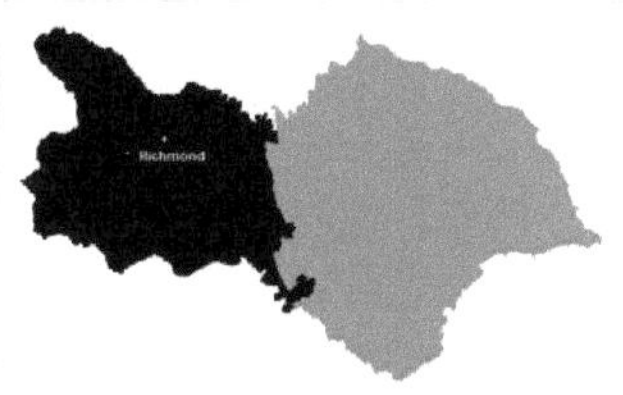

Traditional Richmondshire

One of the most distinctive Christian names of Richmondshire folk is Marmaduke, but this is an old and fading tradition.

Ecclesiastical divisions

St. Paulinus baptised the locals in the River Swale[9] and as a result, it was known as the "Jordan of England".[10] Richmondshire is an Archdeaconry which historically consisted not only of present-day Richmondshire, but also the Barony of Kendal in Westmorland, Copeland (borough) in Cumberland and what is presently Lancashire north of Ribblesdale, such as Amounderness and Lonsdale (hundred).[11] After originally composing part of the Diocese of York, it was transferred to the Diocese of Chester, before moving into the Diocese of Ripon and Leeds.

Modern history

The current district was formed on 1 April 1974, under the Local Government Act 1972. It was a merger of the municipal borough of Richmond with the Aysgarth Rural District, Leyburn Rural District, Reeth Rural District and Richmond Rural District along with part of the Croft Rural District, all in the North Riding of Yorkshire.

The Council was controlled by Independent councillors until May 2003, when elections returned a council with no overall control (Conservative 11; Independent 9; Lib Dem 8; Richmondshire Independent Group 5; Social Democrat Party 1). Conservative John Blackie was elected as leader of the Council. In December 2005 Blackie was replaced by the leader of the Independent Coalition for Richmondshire, Bill Glover, following the resignation of several councillors from the Conservative group and the merger of the rival Independent groups. As of 15 May 2007 the council has been controlled by a minority Conservative administration, with Councillor Melva Steckles as leader of the council.[12]

Settlements

Richmondshire presently includes the major settlements of:

- Askrigg
- Barton
- Catterick
- Catterick Garrison
- Colburn
- Croft-on-Tees
- Hawes
- Keld
- Leyburn
- Middleham
- Middleton Tyas
- Reeth
- Richmond

See also

- Hallamshire
- Hexhamshire
- History of Yorkshire
- Hullshire
- List of hundreds of England and Wales
- Winchcombeshire

References

[1] "Resident Population Estimates by Ethnic Group (Percentages); Mid-2005 Population Estimates" (http://www.neighbourhood.statistics.gov.uk/dissemination/LeadTableView.do?a=3&b=277072&c=richmondshire&d=13&e=13&g=476322&i=1001x1003x1004&m=0&r=1&s=1206654876921&enc=1&dsFamilyId=1812). *National Statistics Online*. Office for National Statistics. . Retrieved 2008-03-28.

[2] http://www.richmondshire.gov.uk/

[3] http://www.richmondshire.gov.uk/about-richmondshire.aspx

[4] "Timeline of the Early British Kingdoms 410 AD-598 AD" (http://www.britannia.com/history/ebk/ebktime1.html). Britannia Internet Magazine. . Retrieved 2009-08-10.

[5] Page (editor), William (1914). "'The honour and castle of Richmond', A History of the County of York North Riding: Volume 1" (http://www.british-history.ac.uk/report.aspx?compid=64709). . Retrieved 2009-08-10.

[6] "The Northern Echo: Bobby Robson, News, Sport, Business, Leisure from the North East and North Yorks - By any name, June brings summer" (http://archive.thenorthernecho.co.uk/2003/6/6/89546.html). archive.thenorthernecho.co.uk. . Retrieved 2009-08-10.

[7] "Richmondshire - Introduction" (http://www.british-history.ac.uk/report.aspx?compid=64710). www.british-history.ac.uk. . Retrieved 2009-08-10.

[8] Hey, David (2005). "3". *A History of Yorkshire*. Lancaster: Garnegie. pp. 88–90. ISBN 10:1-85936-122-6.

[9] "Britannia Biographies: St. Paulinus, Archbishop of York" (http://www.britannia.com/bios/abofy/paulinus.html). Britannia.com. . Retrieved 2009-08-10.

[10] "Yorkshire legends and traditions - Google Books" (http://books.google.co.uk/books?id=vyKiIKB8R9AC&pg=PA7&lpg=PA7&dq="Jordan+of+England"&source=bl&ots=aB0jsiKEMj&sig=wSqHQfYuDP_r-mUhofvwbTpc1N8&hl=en&ei=-2WASpi4EJrLjAe235jxAQ&sa=X&oi=book_result&ct=result&resnum=2#v=onepage&q="Jordan of England"&f=false). books.google.co.uk. . Retrieved 2009-08-10.

[11] "Probate, Lancashire genealogy" (http://www.genuki.org.uk/big/eng/LAN/probate.html). Genuki. . Retrieved 2009-08-10.

[12] "New Leader for Council" (http://www.richmondshire.gov.uk/Default.aspx?page=30407). Richmondshire District Council. 7 January 2008. . Retrieved 2009-08-04.

Bibliography

- *The Pilgrimage of Grace: The rebellion that shook Henry VIII's throne* by Geoffrey Moorhouse
- *The Wars of the Roses* by John Gillingham
- *The Pilgrimage of Grace: and the politics of the 1530s* by R. W. Hoyle.
- *The Penguin Illustrated History of Britain and Ireland: from earliest times to the present day* by Barry Cunliffe, Robert Bartlett (historian), John Morrill (historian), Asa Briggs and Joanna Burke
- *The Swale: A history of the Holy River of St Paulinus* by David Morris
- *The Honour of Richmond: a history of the lords, earls and dukes of Richmond* by David Morris
- *Conquest, Anarchy & Lordship: Yorkshire, 1066-1154* by Paul Dalton
- *Albion's Seed: Four British Folkways in America* by David Hackett Fischer
- *Yorkshire Dales* by Ron Scholes
- *'Richmondshire: Introduction', A History of the County of York North Riding: Volume 1* (http://www.british-history.ac.uk/report.aspx?compid=64710) by William Page
- *Richmond: Geographical and Historical information from the year 1890* (http://www.genuki.org.uk/big/eng/YKS/NRY/Richmond/Richmond90.html) from Bulmer's History and Directory of North Yorkshire (1890)
- *The Early History of Bedale* by H. B. McCall
- *BY ANY NAME, JUNE BRINGS SUMMER* (http://archive.thenorthernecho.co.uk/2003/6/6/89546.html) by The Northern Echo
- A Dictionary of First Names, Oxford University Press, ISBN 0192800507 (http://www.ancestry.com/facts/Marmion-family-history.ashx?fn=Marmaduke&Submit=Check+meaning)

North_Yorkshire

North Yorkshire	
Geography	
Status	Ceremonial & (smaller) Non-metropolitan county
Origin	1974
Region	Yorkshire and the Humber (Part of the ceremonial county is in the North East Region)
Area **- Total**	Ranked 1st 8654 km^2 (**unknown operator: u'strong'** sq mi)
Admin HQ	Northallerton[1]
ISO 3166-2	GB-NYK
ONS code	36
NUTS 3	UKE22
Demography	
Population **- Total (2005)** **- Density** **- Admin. council** **- Admin. pop.**	Ranked 16th 1,082,000 126 /km^2 (**unknown operator: u'strong'** /sq mi) Ranked 19th 599,800
Ethnicity	97.9% White 1.0% S.Asian 1.1% Other
Politics	

North Yorkshire County Council
http://www.northyorks.gov.uk/

Executive	Conservative
Members of Parliament	• Nigel Adams (C) • Hugh Bayley (L) • Stuart Bell (L) • Tom Blenkinsop (L) • Robert Goodwill (C) • William Hague (C) • Andrew Jones (C) • Anne McIntosh (C) • Julian Smith (C) • Julian Sturdy (C) • Ian Swales (LD) • James Wharton (C)
Districts	
1.	Selby
2.	Harrogate
3.	Craven
4.	Richmondshire
5.	Hambleton
6.	Ryedale
7.	Scarborough
8.	City of York (Unitary)
9.	Redcar and Cleveland (Unitary)
10.	Middlesbrough (Unitary)
11.	Stockton-on-Tees (Unitary) (the part south of the Tees)
Neighbouring counties are: County Durham, East Riding of Yorkshire, Cumbria, Lancashire, South Yorkshire and West Yorkshire	

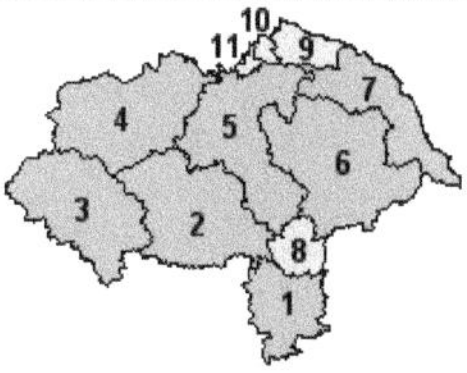

North Yorkshire is a county in England. It is a non-metropolitan or shire county located in the Yorkshire and the Humber region of England, and a ceremonial county primarily in that region but partly in North East England. Created in 1974 by the Local Government Act 1972[2] it covers an area of 8654 square kilometres (**unknown operator: u'strong'** sq mi), making it the largest ceremonial county in England. The majority of the Yorkshire Dales and all of the North York Moors lie within North Yorkshire's boundaries, and around 40% of the county is covered by National Parks. The county town is Northallerton.

Divisions and environs

The area under the control of the county council, or shire county, is divided into a number of local government districts; they are Craven, Hambleton, Harrogate, Richmondshire, Ryedale, Scarborough and Selby.[3]

The Department for Communities and Local Government did consider reorganising North Yorkshire County Council's administrative structure by abolishing the seven district councils and the county council to create a North Yorkshire unitary authority. The changes were planned to be implemented no later than 1 April 2009.[4] [5] This was rejected on 25 July 2007 so the County Council and District Council structure will remain.[6]

The largest settlement in the administrative county is Harrogate, while in the ceremonial county it is York.

York, Middlesbrough and Redcar and Cleveland are unitary authority boroughs which form part of the ceremonial county for various functions such as the Lord Lieutenant of North Yorkshire, but do not come under county council control. Uniquely for a district in England, Stockton-on-Tees is split between North Yorkshire and County Durham for this purpose. Middlesbrough, Stockton-on-Tees, and Redcar and Cleveland boroughs form part of the North East England region.[7]

The ceremonial county area, including the unitary authorities, borders East Riding of Yorkshire, South Yorkshire, West Yorkshire, Lancashire, Cumbria and County Durham.

Physical features

The geology of North Yorkshire is closely reflected in its landscape. Within the county are the North York Moors and most of the Yorkshire Dales; two of eleven areas of countryside within England and Wales to be officially designated as national parks. Between the North York Moors in the east and the Pennine Hills in the west lie the Vales of Mowbray and York. The Tees Lowlands lie to the north of the North York Moors and the Vale of Pickering lies to the south. Its eastern border is the North sea coast. The highest point is Whernside, on the Cumbrian border, at 736 metres (**unknown operator: u'strong'** ft).[8] The three major rivers in the county are the River Swale, River Ure and the River Tees. The Swale and the Ure form the River Ouse which flows through York and into the Humber estuary. The Tees forms the border between North Yorkshire and County Durham and flows from upper Teesdale to Middlesbrough and Stockton and to the coast.

History

North Yorkshire was formed on 1 April 1974 as a result of the Local Government Act 1972, and covers most of the lands of the historic North Riding, as well as the northern half of the West Riding, the northern and eastern fringes of the East Riding of Yorkshire and the former county borough of York.

York became a unitary authority independent of North Yorkshire on 1 April 1996,[9] and at the same time Middlesbrough, Redcar and Cleveland and areas of Stockton-on-Tees south of the river became part of North Yorkshire for ceremonial purposes, having been part of Cleveland from 1974 to 1996.

Governance

North Yorkshire is a non-metropolitan county that operates a cabinet-style council.[10] The full council of 72 elects a council leader, who in turn appoints up to 9 more councillors to form the executive cabinet. The cabinet is responsible for making decisions in the County. The county council have their offices in the county hall in Northallerton.

Economy

Agriculture is an important industry, as are mineral extraction and power generation. The county also has healthy high technology, service and tourism sectors.[11] The number of people claiming Job Seeker's Allowance is low, at 2.7%.[12]

This is a chart of trend of regional gross value added for North Yorkshire at current basic prices with figures in millions of British Pounds Sterling.[13]

Year	Regional Gross Value Added[14]	Agriculture[15]	Industry[16]	Services[17]
1995	**7,278**	478	2,181	4,618
2000	**9,570**	354	2,549	6,667
2003	**11,695**	390	3,025	8,281

Education

North Yorkshire LEA has a mostly comprehensive education system with 42 state schools secondary (not including sixth form colleges) and 12 independent schools.

Climate

North Yorkshire has a Temperate Climate like much of the UK. However there are large climate variations within the county. The upper Pennines border on a Subarctic climate, whereas the Vale of Mowbray has an almost Semi-arid climate. Overall, with the county being situated in the east, it receives below average rainfall for the UK, but the upper Dales of the Pennines are one of the wettest parts of England, where in contrast the driest parts of the Vale of Mowbray are some of the driest areas in the UK. Summer temperatures are above average, at 22°C, but highs can regularly reach up to 28°C, with over 30°C reached in Heatwaves. Winter temperatures are below average, with average lows of 1°C. Snow and Fog can be expected depending on location, with the North Yorkshire Moors and Pennines having snow lying for an average of between 45 and 75 days per year.[18] Sunshine is most plentiful on the coast, receiving an average of 1650 hours a year, and reduces further west in the county, with the Pennines only receiving 1250 hours a year.

Towns and villages

Settlements by population

Rank	Town	Population	Year	Borough	Definition	Notes
1	Harrogate	71,594	2001	Harrogate	Town	Unparished; collection of wards
2	Scarborough	50,135	2001	Scarborough	Town	Unparished; collection of wards
3	Ripon	15,922	2001	Harrogate	Civil Parish	
4	Northallerton	15,741	2001	Hambleton	Civil Parish	
5	Knaresborough	14,740	2001	Harrogate	Civil Parish	
6	Skipton	14,313	2001	Craven	Civil Parish	
7	Whitby	13,594	2001	Scarborough	Civil Parish	
8	Selby	13,012	2001	Selby	Civil Parish	
9	Richmond	8,178	2001	Richmondshire	Civil Parish	
10	Tadcaster	7,000	2001	Selby	Civil Parish	
11	Norton	6,943	2001	Ryedale	Civil Parish	
12	Pickering	6,846	2001	Ryedale	Civil Parish	
13	Filey	6,819	2001	Scarborough	Civil Parish	
14	Sherburn-in-Elmet	6,221	2001	Selby	Civil Parish	
15	Killinghall	5,230[19]	2010	Harrogate	Civil Parish	

Italicised locations lie outside the current North Yorkshire shire county.

- Ampleforth, Appleton-le-Moors
- Bedale, Bolton-on-Swale, Boroughbridge, Borrowby (Hambleton), Borrowby (Scarborough), Brompton (Hambleton), *Brotton*, Buckden
- Castleton, Catterick, Catterick Garrison, Cawood, Clapham, Conistone
- Dalton (Hambleton), Dalton (Richmondshire), Danby Wiske
- Easby (Hambleton), Easingwold, Egton, *Eston*, Ebberston
- Filey, Folkton, Flixton, North Yorkshire
- Giggleswick, Glasshouses, Goathland, *Grangetown*, Grassington, Great Ayton, Grosmont, *Guisborough*, Ganton
- Harrogate, Hawes, Hebden, Helmsley, High Bentham, Horton in Ribblesdale, Hunmanby, *Huntington*
- Ingleton
- Kettlewell, Kilnsey, Kirkbymoorside, Knaresborough
- Leyburn
- Malham, Malton, Masham, *Marske-by-the-Sea*, Middleham, *Middlesbrough*, Middleton, Ryedale, Muston
- *New Marske*, *Normanby*, Northallerton, Norton-on-Derwent, North Grimston,
- Osmotherley, *Ormesby*
- Pateley Bridge, Pickering
- *Redcar*, Reeth, Richmond, Rievaulx, Ripon, Robin Hood's Bay, Romanby
- *Saltburn*, Scarborough, Scorton, Selby, Settle, Sherburn-in-Elmet, Sheriff Hutton, Skelton, Skipton, *South Bank*, Sowerby, Stokesley, Streetlam, Sutton-under-Whitestonecliffe, Swinton, near Malton, Scagglethorpe, Scampston,
- Tadcaster, *Teesville*, Thirsk, *Thornaby-on-Tees*
- *Whale Hill*, Whitby, Westow, Wintringham
- *Yarm*, *York*, Yedingham

Places of interest

Italicised locations lie outside the current North Yorkshire shire county.

Bolton Abbey

Knaresborough

- Ampleforth College
- Beningbrough Hall
- Bolton Abbey
- Bolton Castle
- Brimham Rocks
- Byland Abbey - English Heritage (EH)
- Castle Howard and the Howardian Hills
- Catterick Garrison
- Cleveland Hills
- Drax
- Duncombe Park - stately home
- Embsay & Bolton Abbey Steam Railway
- Eston Nab
- Falconry
- Flamingo Land Theme Park and Zoo
- Fountains Abbey
- *Gisborough Priory*
- Helmsley Castle - EH
- Ingleborough Cave - show cave
- Kirkham Priory

- Lightwater Valley
- Malham Cove
- Middleham Castle
- Mount Grace Priory - EH
- North Yorkshire Moors Railway
- Ormesby Hall - Palladian Mansion
- Richmond Castle
- Rievaulx Abbey - EH

An ancient derelict hunting lodge in Dob Park, North Yorkshire.

- Ripley Castle - Stately home and historic village
- Ripon Cathedral
- Selby Abbey
- Scarborough Castle - EH
- Shandy Hall - stately home
- Skipton Castle
- Stanwick Iron Age Fortifications - EH
- Studley Royal Park
- Stump Cross Caverns - show cave
- Thornborough Henges
- Wharram Percy
- Whitby Abbey
- *York Minster*
- *Yorkshire Air Museum*

News and media

The County is served by BBC North East and Cumbria, and for more southerly parts of the county BBC Yorkshire. Yorkshire Television and Tyne Tees Television are also received in most areas of the County. BBC Tees is broadcast to northern parts of the county, whist BBC Radio York is broadcast more widely. BBC Radio Leeds is broadcast to southern parts of the county.

Transport

The main north-south road through the county is the A1/A1(M) which has gradually been upgraded to motorway status since the early 1990s. The only other motorways within the county are the short A66(M) near Darlington and a small stretch of the M62 motorway close to Eggborough.[3] The other nationally maintained trunk routes are the A168/A19, A64, the A66 and A174.

The East Coast Main Line (ECML) bisects the county stopping at Northallerton, Thirsk and York. Passenger services on the ECML within the county are operated by East Coast, First TransPennine Express and Grand Central. First TransPennine Express run services on the York to Scarborough Line and the Northallerton–Eaglescliffe Line (for Middlesbrough) that both branch off the ECML.

Northern Rail operate the remaining lines in the county including commuter services on the Harrogate Line, Airedale Line and York & Selby Lines, of which the former two are covered by the Metro ticketing area. Remaining branch lines operated by Northern include the Yorkshire Coast Line from Scarborough to Hull, the Hull to York Line via Selby, the Tees Valley Line from Darlington to Saltburn and the Esk Valley Line from Middlesbrough to Whitby. Last but certainly not least, the Settle-Carlisle Line runs through the west of the county with services again operated by Northern.

The county suffered badly under the Beeching cuts of the 1960s which saw places like Richmond, Ripon, Tadcaster, Helmsley, Pickering and the Wensleydale communities lose their passenger services. Notable lines closed were the Scarborough and Whitby Railway, Malton and Driffield Railway and the secondary main line between Northallerton and Harrogate via Ripon.

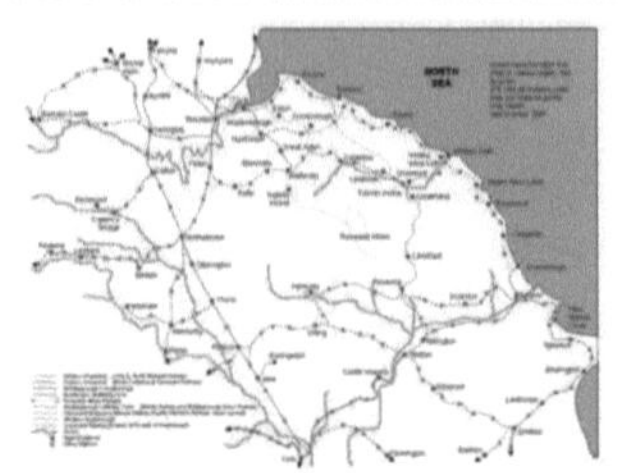
Current and former railway routes in eastern North Yorkshire

Heritage railways within North Yorkshire include the North Yorkshire Moors Railway between Pickering and Grosmont, which opened in 1973, the Derwent Valley Light Railway near York, and the Embsay and Bolton Abbey Steam Railway. The Wensleydale Railway, which started operating in 2003, runs services between Leeming Bar and Redmire along a former freight only line. The medium term aim to operate into Northallerton station on the ECML once agreement can be reached with Network Rail, whilst the longer term aim to re-instate the full line west via Hawes to Garsdale on the Settle-Carlisle line.

York railway station is the largest station in the county with 11 platforms and is a major tourist attraction in its own right. The station is immediately adjacent to the world famous National Railway Museum.

As well as long distance coach services operated by National Express and Megabus local bus service operators include Arriva, Harrogate & District, Scarborough & District (East Yorkshire Motor Services), Yorkshire Coastliner, First and the local Dales & District.

There are no major airports in the county itself but nearby airports include Durham Tees Valley, Newcastle, Robin Hood Doncaster Sheffield and Leeds Bradford.

Sports

North Yorkshire is home to several football clubs, the most successful of which is Middlesbrough FC who play in the Coca-Cola Championship; others include York City FC who have played in the Football League but today play in the Conference National. Whitby Town FC have reached the FA cup first round seven times, and have played the likes of Hull City, Wigan and Plymouth Argyle, they currently play in the Evo-Stik Premier league.

No notable rugby union teams hail from the county but York City Knights are a rugby league team and play in the Rugby League Championship.

North Yorkshire is home to many racecourses; these include Catterick Bridge, Redcar, Ripon and Thirsk. It also has one motor racing circuit, Croft Circuit; the circuit holds meetings of the British Touring Car Championship, British Superbike and Pickup Truck Racing race series.

Yorkshire County Cricket Club, play a number of fixtures at North Marine Road, Scarborough.

The ball game Rock-It-Ball was developed in the county.

See also

- List of Lord Lieutenants of North Yorkshire
- List of High Sheriffs of North Yorkshire

References

[1] "North Yorkshire County Council : Contact us" (http://www.northyorks.gov.uk/index.aspx?articleid=42). www.northyorks.gov.uk. . Retrieved 2009-05-16.

[2] Arnold-Baker, C., *Local Government Act 1972*, (1973)

[3] "Transport map of shire county divided into districts" (http://www.northyorks.gov.uk/CHttpHandler.ashx?id=514&p=0) (PDF). North Yorkshire County Council. . Retrieved 2008-10-10.

[4] "New council for North Yorkshire" (http://www.northyorks.gov.uk/public/site/NYCC/menuitem.72980bf1db3dfb9fd7428f1040008a0c/?vgnextoid=cfa68f0788110110VgnVCM100000420f1cacRCRD). North Yorkshire County Council. .

[5] "Proposals for future unitary structures: Stakeholder consultation" (http://www.communities.gov.uk/documents/localgovernment/pdf/322770.pdf) (PDF). Communities and Local Government. . Retrieved 2008-10-10.

[6] "Decision letter: North Yorkshire County Council" (http://www.communities.gov.uk/pub/59/DecisionletterNorthYorkshireCountyCouncil_id1512059.pdf) (PDF). Communities and Local Government. .

[7] North East Assembly (http://www.northeastassembly.gov.uk/theassembly/list.cfm) - List of local authorities and members

[8] http://bubl.ac.uk/org/tacit/marilyns/chapter6.htm

[9] OPSI (http://www.opsi.gov.uk/SI/si1995/Uksi_19950610_en_1.htm#end) - *The North Yorkshire (District of York) (Structural and Boundary Changes) Order 1995*

[10] "North Yorkshire County Council Constitution" (http://www.northyorks.gov.uk/CHttpHandler.ashx?id=1990&p=0). North Yorkshire County Council. . Retrieved 10 May 2010.

[11] http://www.northyorks.gov.uk/index.aspx?articleid=2874

[12] http://www.northyorks.gov.uk/CHttpHandler.ashx?id=12525&p=0

[13] "Regional Gross Value Added" (http://www.statistics.gov.uk/downloads/theme_economy/RegionalGVA.pdf) (PDF). Office for National Statistics. 2005-12-21. pp. 240–253. . Retrieved 2008-10-06.

[14] Components may not sum to totals due to rounding

[15] includes hunting and forestry

[16] includes energy and construction

[17] includes financial intermediation services indirectly measured

[18] "Regional mapped climate averages" (http://www.metoffice.gov.uk/climate/uk/averages/regmapavge.html#). The Met Office. . Retrieved 27 September 2010.

[19] http://www.northyorks.gov.uk/CHttpHandler.ashx?id=16424&p=0

External links

- North Yorkshire Guide (http://www.northyorks.com) Guide from NorthYorks.com
- BBC North Yorkshire (http://www.bbc.co.uk/northyorkshire/) North Yorkshire features, videos and pictures from the BBC
- Yorkshire Dales Rivers Trust (http://www.yorkshiredalesriverstrust.org.uk/)
- Discover North Yorkshire (http://www.discovernorthyorkshire.co.uk) North Yorkshire Tourist Information
- North Yorkshire (http://www.dmoz.org/Regional/Europe/United_Kingdom/England/North_Yorkshire/) at the Open Directory Project

Reeth

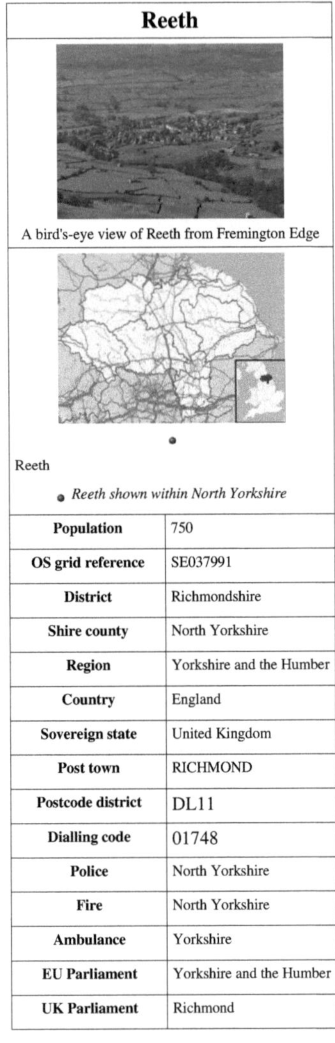

Reeth

A bird's-eye view of Reeth from Fremington Edge

Reeth

Reeth shown within North Yorkshire

Population	750
OS grid reference	SE037991
District	Richmondshire
Shire county	North Yorkshire
Region	Yorkshire and the Humber
Country	England
Sovereign state	United Kingdom
Post town	RICHMOND
Postcode district	DL11
Dialling code	01748
Police	North Yorkshire
Fire	North Yorkshire
Ambulance	Yorkshire
EU Parliament	Yorkshire and the Humber
UK Parliament	Richmond

Reeth is a village in the Yorkshire Dales within the Richmondshire district of North Yorkshire, England and principal settlement of Swaledale.[1] It is situated at the meeting point of the two most northerly of the Yorkshire Dales: Swaledale and Arkengarthdale.

In Saxon times, Reeth was only a settlement on the forest edge, but by the time of the Norman Conquest it had grown sufficiently in importance to be noted in the Domesday Book.

Later it became a centre for hand-knitting and the local lead industry was controlled from here, but it was always a market centre for the local farming community. Its 18th-century houses and hotels are clustered around a triangular green. The village has three pubs all situated on the green. They are the Black Bull, the King's Arms and the Buck Hotel. The village is overlooked by the fells of Fremington Edge and Calver Hill. The village also offers several B & Bs and tea rooms as well as a hotel, the Burgoyne, that overlooks the village green.

The village, known as the capital of Swaledale, is the starting point for several of the North Yorkshire Dales' best known walks. Most notable is Fremington Edge, a walk that, after a steady climb, provides good views of Swaledale.

The Village Square in Reeth.

In May and June every year, Reeth becomes the hub of the Swaledale Festival, a two-week celebration of small-scale music and guided walks. Additionally on the final Wednesday of August, the Reeth Show, an agricultural event, is held.[2] In 2012 the show will enter its centenary year .

Notable people

- The nonconformist minister David Bradberry was born in Reeth.
- Ruby Ferguson, a writer of popular fiction including children's books, romances, and mysteries, was born and raised in Reeth.

See also

- Swaledale Museum in Reeth

References

[1] Reeth, Yorkshire Dales (http://www.chromavision.co.uk/yt/reeth.htm), Yorkshire Tour, UK.
[2] Reeth Show (http://www.reethshow.co.uk/), UK.

External links

- Reeth Information from Reeth.org (http://www.reeth.org)
- Reeth Guide (http://www.northyorks.com/reeth.htm)

Telephone_booth

A **telephone booth**, **telephone kiosk**, **telephone call box** or **telephone box** is a small structure furnished with a payphone and designed for a telephone user's convenience. In the USA, Canada and Australia, "telephone booth" is used, while in the UK and the rest of the Commonwealth it is a "telephone box" or "phone box". Such a booth usually has a door to provide privacy and a window to let others know if the booth is in use. The booth may be furnished with a printed directory of local telephone numbers, and a booth in a formal setting such as a hotel may be furnished with paper and pen and even a seat. An outdoor booth may be made of metal and plastic to withstand the elements and heavy use, while an indoor booth (once known as a silence cabinet) may have more elaborate architecture and furnishings.[1] Most outdoor booths feature the name and logo of the telephone service provider.

Classic UK red telephone boxes.

A K6 red telephone box and a Edward VII pillar box at the Amberley Working Museum.

Design

A example of a person using a Telstra phone box in Victoria, Australia

Starting in the 1970s pay telephones were less and less commonly placed in booths in the United States. In many areas where they were once common, telephone booths have now been almost completely replaced by non-enclosed pay phones. In the United States, this replacement was caused, at least in part, by an attempt to make the pay telephones more accessible to disabled people. However, in the United Kingdom telephones remained in booths more often than the non-enclosed set up. Although still fairly common, the number of phone boxes has declined sharply in Britain since the late 1990s due to the boom of mobile phones.

Many locations that provide pay-phones mount the phones on kiosks rather than in booths — this relative lack of privacy and comfort discourages lengthy calls in high-demand areas such as airports.

Special equipment installed in some telephone booths allows a caller to use a computer, a portable fax machine, or a telecommunications device for the deaf.

Paying for the call

Coins

The user of the booth pays for the call by depositing coins into a slot on the telephone. Coin-operated phones usually take the money before the call is made, and return it if there is no answer on the receiving end. Other phones, such as those used in the UK until the early 1980s, take the money after someone has answered at the other end by blocking the call until money has been deposited.

Cards

Calls may be paid for by entering a payment code on the telephone's keypad, by swipe-card ("*Swipe & Call*") or by using a telephone card. Some pay phones are equipped with a card reader that allows a caller to make payment with a credit card.

Collect call

A caller who possesses no means of payment may have the phone company's operator ask the call recipient if the recipient is willing to make payment for the call, this is known as "reversing the charges", "reverse charged call" or a "collect call". It is also possible to place a call to a phone booth if the intended recipient is known to be waiting at the booth, however not all phone booths allow incoming calls. Long before "computer hacking" was a common phenomenon, creative mischief-makers devised tactics for obtaining free phone usage through a variety of techniques, including several for defeating the electro-mechanical payment mechanisms of telephone booths—early methods of phone phreaking.

History

The first telephone booth was probably located near the Staple Inn in High Holborn(London) in May of 1903.[2] It was operated and located by the Grand Central Railway. However, some sources claim that there was a telephone box called "Fernsprechkiosk" in Berlin in 1881.

Recent developments

Wireless services

Phone boxes are often vandalised

The increasing use of mobile phones has led to a decreased demand for pay telephones, but the increasing use of laptops is leading to a new kind of service. In 2003, service provider Verizon announced that they would begin offering wireless computer connectivity in the vicinity of their phone booths in Manhattan. As of 2006 the Verizon wifi telephone booth service was discontinued in favor of the more expensive Verizon Wireless's EVDO system.[3] This allows a computer user to connect with remote computer services by means of a short-range radio stationed within the booth. The caller pays for usage by means of a pre-arranged account code stored inside the caller's computer. Wireless access is motivating telephone companies to place wireless stations at locations that have traditionally hosted telephone booths, but stations are also appearing in new kinds of locations such as libraries, cafés, and trains.

Vandalism

Calling cards are often found in phone boxes in London advertising the services of call girls

A rise in vandalism in certain regions has prompted several companies to manufacture simpler booths with extremely strong pay-phones.

Dual currencies

Most telephone booths in Northern Ireland are able to accept two currencies. They are able to accept both pound sterling and euro, due to the proximity to the Republic of Ireland. Similarly, mainly in large cities in Great Britain, certain telephone booths accept both sterling and euro. Other services provided by these booths are internet access, SMS text messaging and ordinary phone services.

Withdrawal of services

Pay-phones still may get used by mobile phones users if their phone breaks, runs out of battery or gets stolen, or for other emergency uses, which may makes a complete disappearance in the near future less likely.

Jordan

However, in 2004, Jordan became the first country in the world not to have telephone booths generally available. The cellular phone penetration in that country is so high that telephone booths have practically not been used at all for years. The two private payphone service companies, namely ALO and JPP, closed down and currently there's no payphone service to speak of.[4]

Finland

In 2007, Finnet companies, and TeliaSonera Finland discontinued their public telephones earlier and the last remaining operator Elisa Oyj did that during the beginning of the year.[5]

A telephone booth in Brazil, popularly called *orelhão* ("big ear") because of its format.

Smoking ban

Following the commencement of the Smoking ban in England, it became illegal to smoke in most telephone boxes. The smoking ban requires owners to display no smoking signage, which has resulted in BT displaying a no smoking sticker which refer to the telephone box as "premises".

Charging points

Since many telephone boxes tend to be at the roadside and already have electricity supplies, a trial is to take place in the Spanish capital, Madrid, to convert 30 former telephone boxes into charging points for electric cars.[6]

A joke public telephone at a V-Day anniversary celebration in Hampshire.

Advertising

Many telephone boxes in the UK have became points of advertisement, bearing posters, with the development of StreetTalk by JCDecaux.[7]

See also

- Callbox
- Giles Gilbert Scott
- Interactive kiosk
- Mojave phone booth
- Payphone
- Police Box
- Red telephone box

References

[1] "Public Telephones" (http://www.melchior.co.uk/BTphones/phone2.html). Melchior Telematics. . Retrieved December 4, 2007.

[2] http://pl.wikipedia.org/wiki/Budka_telefoniczna

[3] Jen Chung (May 2, 2005). "Goodbye Free Verizon WiFi" (http://gothamist.com/2005/05/02/goodbye_free_verizon_wifi.php). Gothamist LLC. . Retrieved December 4, 2007.

[4] "Payphones suffer from cellphone growth 2004" (http://www.cellular.co.za/news_2004/march/032704-payphones_suffer_from_cellphone.htm). CellularOnline. March 22 2004. .

[5] "Elisa luopuu yleisöpuhelinliiketoiminnasta syksyllä 2007" (http://www.elisa.fi/ir/index.cfm?t=5&o=5120.00&did=13602) (in Finnish). Elisa Oyj. November 15, 2006. . Retrieved December 4, 2007.
[6] Tremlett, Giles (September 8, 2009). "Madrid reverses the chargers with electric car plan" (http://www.guardian.co.uk/environment/2009/sep/08/electric-car-plan-spain). The Guardian. .
[7] "JCDecaux StreetTalk" (http://www.jcdecaux.co.uk/products/streettalk/). JCDecaux. . Retrieved September 22, 2011.

External links

- PayPhoneBox (http://www.payphonebox.com/) Index of payphone numbers and photographs of payphones in unusual or famous places around the world.
- *La Cabina* (http://www.imdb.com/title/tt0065513/) at the Internet Movie Database
- Article in Spaces Magazine by Lee Garland, May 08 (http://www.leegarlandphotography.co.uk/Architecture/23 End Space.pdf)
- Starting the revolution of Phone Booths (http://www.framery.fi)

Tadcaster

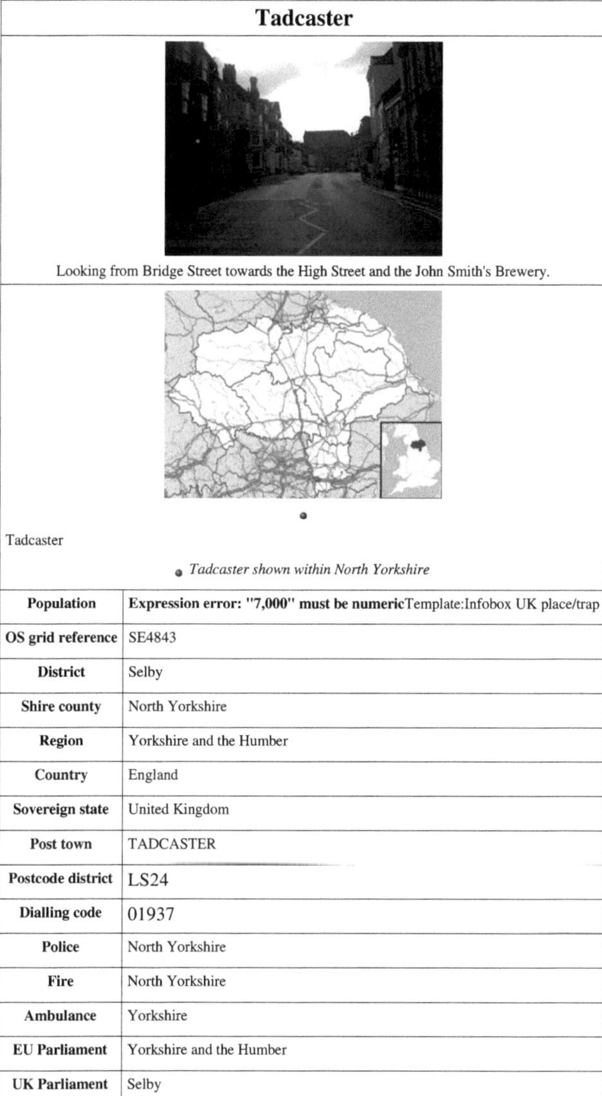

Tadcaster

Looking from Bridge Street towards the High Street and the John Smith's Brewery.

Tadcaster

Tadcaster shown within North Yorkshire

Population	**Expression error: "7,000" must be numeric**Template:Infobox UK place/trap
OS grid reference	SE4843
District	Selby
Shire county	North Yorkshire
Region	Yorkshire and the Humber
Country	England
Sovereign state	United Kingdom
Post town	TADCASTER
Postcode district	LS24
Dialling code	01937
Police	North Yorkshire
Fire	North Yorkshire
Ambulance	Yorkshire
EU Parliament	Yorkshire and the Humber
UK Parliament	Selby

Tadcaster is a market town and civil parish in the Selby district of North Yorkshire, England. Lying near the Great North Road approximately 15 miles (**unknown operator: u'strong'** km) east of Leeds and 10 miles (**unknown operator: u'strong'** km) west of York. It is the last town on the River Wharfe before it joins the River Ouse about 10 miles (**unknown operator: u'strong'** km) downstream. It is part of the shire county of North Yorkshire, despite being further south than York, the traditional centre of Yorkshire and thus historically in the West Riding.

The town is twinned with Saint Chély d'Apcher in France.

Government

Tadcaster Social Club

For local government purposes, the River Wharfe divides the town into eastern and western electoral wards. The combined population of Tadcaster East and Tadcaster West in 2004 was 7,280, 3,800 in Tadcaster East and 3,480 in Tadcaster West (source: Office of National Statistics). The local authority is Selby District Council.

Tadcaster gave its name to a much larger rural district council, Tadcaster Rural District and other administrative areas. This may lead to confusion when comparing the size and extent of the current town with information for earlier periods. For example the population in 1911 of the Tadcaster sub-district was 6831 compared with that of the Tadcaster Registration District, 32052 (source: A Vision of Britain through time).

History

Roman times

Tadcaster was founded by the Romans, who named it *Calcaria* from the Latin word for *lime*, reflecting the importance of the area's limestone geology as a natural resource for quarrying, an industry which continues today and has contributed to many notable buildings including York Minster. Calcaria was an important staging post on the road to Eboracum (York), which grew up at the river crossing.

The Calcaria public house, the namesake of Calcaria

Anglo-Saxon and medieval times

The suffix of the Anglo-Saxon name *Tadcaster* is derived from the borrowed Latin word *castra* meaning 'fort', although the Angles and Saxons used it for any walled Roman settlement. Tadcaster is mentioned in the Anglo-Saxon Chronicle as the place were King Harold assembled his army and fleet prior to entry into York and subsequently on to the Battle of Stamford Bridge in 1066.

The town is mentioned in the 1086 Domesday Book as **Tatecastre**. The record reads: *Two Manors. In Tatecastre, Dunstan and Turchil had eight carucates of land for geld, where four ploughs may be. Now, William de Parci has three ploughs and 19 villanes and 11 bordars having four ploughs, and two mills of ten shillings (annual value). Sixteen acres of meadow are there. The whole manors, five quaranteens in length, and five in breadth. In King Edward's time they were worth forty shillings; now one hundred shillings.*

In the 11th century William de Percy established **Tadcaster Castle**, a motte-and-bailey fortress, near the present town centre using stone reclaimed from Roman rubble. The castle was abandoned in the early 12th century, and though briefly re-fortified with cannon emplacements during the Civil War, all that remains is the castle motte. The outline of the long demolished southern bailey still impacts the geography of surrounding streets.

The original river crossing was probably a simple ford near the present site of St Mary's Church, soon followed by a wooden bridge. Around 1240, the first stone bridge was constructed close by, possibly from stone once again reclaimed from the castle.

Civil War

At 11am on Tuesday 7 December 1642 the Battle of Tadcaster, an incident during the English Civil War, took place on and around Tadcaster Bridge between Sir Thomas Fairfax's Parliamentarian forces and the Sir Thomans Glemham Royalist army. The present day Wharfe bridge was constructed on the foundations of the stone original in around 1700, though it has been substantially modified at least twice since then. Historically, the bridge marks the boundary between the West Riding and the Ainsty of York; important people would have been formally met here on their journey to York.

Market

A market has been held at Tadcaster since 1270, when Henry de Percy obtained a royal charter from Henry III to hold 'a market and fair at his manor of Tadcaster', to be held each Tuesday. This ancient market place can be seen at the junction of Kirkgate and Bridge Street.

A stone base, believed to have been part of the original market cross, used to stand on Westgate, though this position is now held by the Tadcaster War Memorial. The present-day market is held on Thursdays on the carpark of Tadcaster Social Club on St. Josephs Street.

Industry

Tadcaster has long been associated with the brewing industry due to the quality and accessibility of the local water. Rich in lime sulphate after filtering through the Yorkshire limestone, in the right conditions freshwater springs - known locally as *popple-wells* - can still be seen bubbling up near St Mary's church. Tax registers from 1341 record the presence of two thriving breweries or brewhouses in the town, one paying 8d in tax and the other 4d. Today it is second in importance only to Burton-upon-Trent as an English brewing centre.

John Smith's Brewery Tadcaster

Three breweries have survived into the present day, The Tower Brewery (Coors, formerly Bass), John Smith's and Samuel Smith's Old Brewery, which is also the oldest brewery in Yorkshire and the only remaining independent brewery in Tadcaster. A fourth stood by the river on the site of the present central carpark. Sam Smith's dray horses are a common sight on the streets of the town. Tadcaster is one of the very few Yorkshire towns which still has the industry it grew up around. This helps it to maintain a community spirit not often seen in other towns. Having three breweries in the town employing a large number of the locals the breweries have seen fit to subsidise several of the public houses in the town. These establishments offer the locally brewed drinks at approximately half the price you would expect to pay in a city.

Samuel Smith's Old Brewery

Coors brewery (formerly the Bass Brewery) on Wetherby Road

Architecture

The Ark

The oldest building still in active use in Tadcaster is **The Ark**, built in the late 15th century, though it has been enlarged and altered many times since. Two carved heads on the front of the building are thought to represent the heads of Noah and his wife, hence the name. Throughout its life, the Ark has been a meeting place, a post office, an inn, a butcher's, a private house and a museum; it is currently the Town Council offices.

In the 17th century it was known as Morley Hall, and was licensed for Presbyterian meetings. The Pilgrim Fathers met here and are reputed to have planned their voyage to America; an exact replica exists in Ohio.

St Mary's Church

St Mary's Church was first built around 1150, though a wooden structure did exist prior to this. Destroyed by the Scots in 1318 in one of many incursions subsequent to the Battle of Bannockburn, St Mary's was rebuilt between about 1380 and 1480 but constant problems with flooding led to the structure being taken down stone by stone and rebuilt between 1875 and 1877 with the foundations raised by 5 feet (**unknown operator: u'strong'** m); only the tower was left untouched. The money for this renovation - £8,426 4s 6½d - was raised by public subscription. In 1897 a new north aisle was added.

church of St Mary the Virgin.

The Viaduct

Tadcaster viaduct.

A $^{1}/_{4}$-mile (**unknown operator: u'strong'** m) above the Wharfe bridge an imposing viaduct of eleven arches spans the River Wharfe. This was built as part of the projected Leeds to York railway promoted by the industrialist George Hudson through the York and North Midland Railway. The construction of the line was authorised in 1846. It was to run from Copmanthorpe to Cross Gates, joining the Church Fenton to Harrogate railway line between Tadcaster and Stutton. The collapse of railway investment in 1849 lead to the line being abandoned after the viaduct had been constructed. The need for the line evaporated with the opening of the Micklethorpe to Church Fenton line in 1869. Between 1883 and 1959 the viaduct carried a siding that serviced a mill on the East side of the River Wharfe. The last time the viaduct was used to fetch and carry goods was in 1955. The structure is now a grade II listed building owned by Tadcaster Town Council for the use and pleasure of the local people.[1]

Conservation

To the south east of the town centre, towards the village of Oxton, lies **Tadcaster Mere**. Designated as a Site of Special Scientific Interest or SSSI in 1987, the Mere is in fact the central part of a former lake basin which extended over an area of about 3 km². It was formed during the most recent or Devensian ice age (which ended 10,000 years ago, when present-day Tadcaster would have been situated at the southernmost limit of glaciation) by the long, low embankment of debris known as the Escrick Moraine, which is composed of debris left behind by the Vale of York Glacier. The Mere is a site of current archaeological interest, as it is currently believed to be the site of the earliest discovery of the plesiosaur (while unproven, the skeletal fragments found in Tadcaster at the least match the age of those found elsewhere).[2]

Scientific analysis of the mere, in particular sedimentary pollen studies, provides insight into the geological history and makeup of the local environment and allows accurate dating of events before, during and after the Devensian ice age.

Education

Tadcaster has three Primary Schools (serving ages 5–11) and one Secondary School (ages 11–18). In the Summer 1999 League Tables, Tadcaster Grammar School students obtained the best A Level results in the country for a state comprehensive school. There is also an adult education centre, co-located with the Grammar School.

In the past, Tadcaster served the Wetherby/Tadcaster area with a grammar school, while the Secondary Modern was at Wetherby (what is now Wetherby High School). Since Wetherby is now part of the City of Leeds and Tadcaster is part of the District of Selby, it has become difficult to arrange for pupils to be educated on the opposite side of the border line from the one on which they live.

Media

Local newspapers covering Tadcaster include *The Press* and *The Wetherby News*. The major regional newspaper in the area is the *Yorkshire Post*.

The local BBC radio station is Radio York, and commercial stations include Minster FM and Galaxy 105.

Sport

Tadcaster has two main sports rivals within the town; Tadcaster Albion and Tadcaster Magnets. Tadcaster Magnets also have a significant rivalry with the other major local team Ulleskelf.

Another popular group in the area is the Tadcaster Harriers running club which has been going for over 25 years.

Tadcaster Tornadoes Basketball Team play in the Leeds Basketball League (Mens) and are based at the Tadcaster Leisure Centre.

The route of The White Rose Way, a long distance walk from Leeds to Scarborough passes through the town.

Leisure

Leisure Centre

The Leisure Centre on Station Road can be hired throughout the day for a variety of activities that including Badminton, Roller Skating, Basketball, Volleyball, Indoor Cricket, Tennis, Short Tennis. Bookings can be made up to 7 days in advance. Various private sports clubs are run from Tadcaster Leisure Centre (including Tadcaster Tornadoes Basketball Team), and there is a physiotherapy clinic available on-site.

Swimming Pool

The Tadcaster community swimming pool opened in December 1994 and is run as a charity. At the end of 2007 the pool closed temporarily for repairs to the tiles on the pool floor, reopening on 31 May 2008. The townspeople came together to raise the £130,000 needed to repair the tiles and organised many different events including a celebrity football match against the cast of Emmerdale.

The swimming pool has a fitness suite. There are further swimming pools in Wetherby and York, whist the nearest Olympic pool is at the John Charles Centre for Sport in Beeston, Leeds.

Public transport

Tadcaster is well served by local bus services [3] operating from Leeds City bus station. The town is a main stop on the Yorkshire Coastliner service, which provides easy access to the Yorkshire Coast.

The nearest railway stations are in the villages of Ulleskelf and Church Fenton, but the most convenient station is York as it has a much wider range of services and is connected to Tadcaster by a fairly frequent Yorkshire Coastliner bus service running from outside the railway station.

There are also several mini bus companies within Tadcaster.

References

[1] National Monuments Record entry for the viaduct (http://pastscape.english-heritage.org.uk/hob.aspx?hob_id=54977&sort=1&type=&class1=None&period=None&county=None&place=tadcaster&yearfrom=ALL&yearto=ALL&recordsperpage=5&source=text&nmr=&defra=&p=6)
[2] http://www.assemblage.group.shef.ac.uk/3/3nicki.htm.
[3] http://getdown.org.uk/bus/search/tadcaster.htm

External links

- Tadcaster website (http://www.tadcaster-on-line.co.uk)
- Local Tadcaster website (http://www.tadcaster.uk.com)
- Historical site (http://www.tadcaster-ww1-memorials.com/)
- Local scooter club (http://www.tadcaster-stax.co.uk)
- Tadcaster Leisure Centre (http://www.selbyleisure.co.uk/tadcaster.htm)
- Tadcaster Tornadoes Basketball Team (http://www.tad-tornadoes.org.uk/)
- Tadcaster Community Swimming Pool Trust (http://www.tadcasterpool.org.uk)
- Yorkshire Coastliner (http://www.yorkshirecoastliner.co.uk/)
- Office of National Statistics (http://www.neighbourhood.statistics.gov.uk)
- Vision of Britain through the ages: Tadcaster (http://www.visionofbritain.org.uk/unit_page.jsp?u_id=10029557)

Reeth,_Fremington_and_Healaugh

Reeth, Fremington and Healaugh is a civil parish in the Richmondshire district of North Yorkshire, England. It consists of the three villages of Reeth, Fremington and Healaugh.

As of the 2001 census, it had a population of 685.[1]

References

[1] "Parish of Reeth, Fremington and Healaugh" (https://www3.northyorks.gov.uk/census2001/parish_pdf/richmondshire/reeth_fremington_healaugh.pdf). Key Statistics, 2001 Census. . Retrieved 12 October 2010.

Article Sources and Contributors

Healaugh *Source*: http://en.wikipedia.org/w/index.php?title=Healaugh *Contributors*: Bjones, Cacolantern, DINOMAN, Euchiasmus, JHunterJ, Keith D, Kreb, Rjwilmsi, Waacstats, 5 anonymous edits

Swaledale *Source*: http://en.wikipedia.org/w/index.php?title=Swaledale *Contributors*: Achmelvic, Almost-instinct, Cacolantern, CanOfWorms, Cap, Carre, Centrx, Chris the speller, Cnyborg, Delta 51, Deor, Dvyost, Euchiasmus, Focaldepth, Harkey Lodger, Hebrides, JD554, James@hopgrove, Jay1279, Jpbowen, Kaly99, Kbdank71, Keith D, Ken Gallager, M R G WIKI999, Mckeedi1, Mejor Los Indios, Mhockey, NorthOnTop, Nreckert, Oliver Chettle, Oliver Han, Padres Hana, Penrithguy, PigFlu Oink, RJE42, Rodhullandemu, Saga City, Stemonitis, Steven Walling, Susvolans, The Anome, Thruxton, Trilobite, Vaelta, 14 anonymous edits

Civil_parishes_in_England *Source*: http://en.wikipedia.org/w/index.php?title=Civil_parishes_in_England *Contributors*: Academic Challenger, Ad.minster, Adam keller, After Midnight, Alex Law, Altenmann, Andrew Gray, Andrew Gwilliam, Aoeuidhtns, Armindo, Axel-berger, BenShade, Berek, Betacommand, Bevo74, Björn Bornhöft, Breadandcheese, Broxi, Bunnyfool, Cavrdg, Chrisjj, Cmdrjameson, Cnyborg, Cobber17, Cutler, Darryl.matheson, David Kernow, Ddstretch, DocWatson42, Dpaajones, Eddaido, Editstat7, Ehrenkater, Eigenwijze mustang, English Lock, Epbr123, Epolk, Exile, FlagSteward, G-Man, Gazilion, Gem, Gerry Lynch, Grinner, Hairy Dude, Herr Satz, Hu, Hymers2, Imgaril, Iridescent, Iten, JaT, Jan1nad, JeremyA, Jesster79, Jhamez84, Jm2153, Jurema Oliveira, Just H, Jvhertum, Jza84, Kaihsu, Koryakov Yuri, Kudpung, LG02, Laurel Bush, LinguisticDemographer, Lozleader, MBisanz, MITBeaverRocks, MRSC, Mais oui!, Malleus Fatuorum, Man vyi, MapsMan, Mario56m2, Markswan, Materialscientist, Mauls, Mcginnly, Mervyn, Mhockey, Mibblepedia, Michael Hardy, Mocking Bird, Morwen, Neddyseagoon, Noisy, Nricardo, Nyttend, O'Dea, PACJac199, PaddyMatthews, Peterkingiron, Pigsonthewing, Piotrus, Pwqn, Rarelibra, Rat144, Redlentil, Regan123, Rich Farmbrough, Richardguk, Roke, SGBailey, SMP, Saga City, Sardanaphalus, Sjorford, Squids and Chips, Staffelde, Str1977, Supertrooper, Sussexonian, TarmoK, That Guy, From That Show!, The Transhumanist, The flying pasty, Tmol42, Vegaswikian, Vik-Thor, Warren Whyte, Wereon, Woohookitty, Wrelwser43, ZimZalaBim, 51 anonymous edits

Yorkshire_Dales *Source*: http://en.wikipedia.org/w/index.php?title=Yorkshire_Dales *Contributors*: 5iron2, 9eyedeel, Achmelvic, Aitias, Alethe, Andeggs, Angela, Anthony, Aquilina, Arronsj, Asterion, Bfigura's puppy, Blackcoffeenosugar, Blisco, Bob Castle, Brholden, Cambyses, Caroig, Chartierk, Childzy, Chris the speller, Clocker, Cnyborg, Colonies Chris, CommonsDelinker, Coolhawks88, Cowzie, DINOMAN, Dave.Dunford, DaveGorman, Dbfirs, Delta 51, Derek Ross, Design Forte, Duncharris, Eipl01, Enchanter, Euchiasmus, Everyking, Extraordinary, Fefour, Gartland, Geopersona, Goatchurch, GrahamHardy, Grinner, Gwernol, Harkey Lodger, Herbythyme, Hmains, Hugo999, Huw Powell, Ics2s3b, Immanuel Giel, J.delanoy, JD554, James26, Jamsta0, Jeremy Bolwell, Jez uk1, Jllm06, Jpbowen, Jschwa1, Jza84, Kappa, Kbdank71, Keith D, Keith Edkins, Kgajos, Kingturtle, Krollo, Lancsalot, Langcliffe, LordHarris, Loren36, Lupin, MPF, Mark J, Martarius, Mckeedi1, Mhockey, Mick Knapton, MightyJordan, Morwen, NawlinWiki, NigelR, Oliver Chettle, Oliver Han, Olivier, Ozymandias444, Pablo X, PhilHibbs, Philip Trueman, Politepunk, PoonJ, Propaniac, RJE42, RedWolf, Reedy, Rehevkor, Rich Farmbrough, Rimmer1993, Ringolad, Robedob22, Rrenner, Rudowsky, SL93, Saga City, Sassf, SchuminWeb, Seadog365, Self willed, Squash, Stemonitis, StephenDawson, Stephenb, SunCreator, Tagishsimon, Tarquin, Template namespace initialisation script, The cows want their milk back, TheCustomOfLife, TheGrappler, Tim P, TimTay, Tong22, Trogg2, Varlaam, VerruckteDan, WTucker, Wighson, YorkDales, Zydeco joe, Ü, 157 anonymous edits

Richmondshire *Source*: http://en.wikipedia.org/w/index.php?title=Richmondshire *Contributors*: Achmelvic, Akc dubelu, Alai, AxG, Baxter Ramsey, Borderer, BrownHairedGirl, Catterick, Chris the speller, Cnyborg, D6, DWaterson, Dalliance, Decumanus, Farrtj, Frontiersman, G-Man, Ghmyrtle, Grutness, Harkey Lodger, Harris000, IP Address, Jag123, Jandalhandler, Jaraalbe, Jhamez84, Jza84, Keith D, Keith Edkins, King of the North East, Kipperfield, LilHelpa, Loganberry, Lozleader, MRSC, Modulatum, Morwen, Nicke Lilltroll, Oliver Chettle, Orioane, Owain, Paulleake, Plastikspork, QuartierLatin1968, Rhode Islander, Rich Farmbrough, Richmondshire, Rodhullandemu, Shanel, Template namespace initialisation script, TheMadBaron, TheUnforgiven, Tim!, Warofdreams, Winston365, Woohookitty, Yorkshire Phoenix, 22 anonymous edits

North_Yorkshire *Source*: http://en.wikipedia.org/w/index.php?title=North_Yorkshire *Contributors*: 24-7group, 666andy, Acabashi, Acalamari, Achmelvic, Airunp, Al Silonov, Alex.muller, Alsybaby, Anark, Andrew Norman, Angela, Angusmclellan, Anwar saadat, Aquilina, Bazonka, Bduke, Beetstra, Biggleswiki, Bluemoose, Bob Castle, Bodhitree108, Borofan4ever, BrownHairedGirl, CKTerra, Cacolantern, Chanheigeorge, Chris j wood, Chris075, Chrisloader, Citterio, Clio64B, Cnyborg, Colinwealleamns, CommonsDelinker, Computerjoe, DCB4W, Dalliance, DavidCane, Davshul, Dhewison, DinosaursLoveExistence, DisillusionedBitterAndKnackered, Dpaajones, Dr.ongo69, Dricherby, Dthomsen8, Dudesleeper, Edward, Elliskev, Enchanter, EricITOworld, Esprit15d, Falconer4, Faradayplank, Farrtj, Fenners, Flatterworld, Fourohfour, Francish7, G-Man, Gaia Octavia Agrippa, Galoubet, Geopersona, Gilderien, Gilliam, GordyB, Grahamp, Greenshed, Gtothegizzo, Gunnar Larsson, Hahnchen, Harkey Lodger, Harrychown, Harrychown1989, Herbythyme, Heron, Hogyn Lleol, Ilikeeatingwaffles, Immanuel Giel, Iridescent, Isaacwatts, Iulianu, Jamsta0, Jance day, Jcuk, Jhamez84, Joseph Solis in Australia, Jschwa1, Jza84, Keith D, Keith Edkins, Ken Gallager, Kummi, Lancsalot, Langcliffe, Le Creillois, Lightmouse, Lisagosselin, LizardJr8, Lofty, Lozleader, MDCollins, MER-C, MRSC, MRacer, Mais oui!, Marknew, Martarius, Matt Cd, Matt henderson1, Max Naylor, Mckeedi1, Michael Glass, Morwen, MrBloodyMinded, NRTurner, Nadavspi, Nard the Bard, Nate1481, Neepa24, Neilirving, Nevilley, NigelR, Nigelcoates, Nilfanion, NinetyCharacters, Nix D, Nono64, Oliver Chettle, Ollie, Owain, Palica, Paul-L, Paulleake, Pjmc, Plucas58, Pluto8888, Pokemonmann, Prestr1, Quercusrobur, R Lowry, RHaworth, RMBair, RedWolf, Redf0x, Renata, Republica, Richard Harvey, Rocastelo, Rrius, Ryedaletourism, S kitahashi, ST47, Saga City, Scott 243, Shimlad, Sjorford, Snigbrook, Stephenkirkup, Stephenswfc, Suruena, TJBlackwell, TastyPoutine, Template namespace initialisation script, TerriersFan, Tesi1700, The JPS, This is Drew, Timrollpickering, Tom-, Tomd999, Tomtolkien, Tracer.smart, Uncantabrigian, Warofdreams, WeakLemonDrink, Welsh, Wikityke, Wikiuser100, William Avery, Wjfox2005, Xtrememachineuk, YMCTP, Yorkshire Phoenix, ZPM, ZaffaNE, Zzuuzz, ²¹², 205 anonymous edits

Reeth *Source*: http://en.wikipedia.org/w/index.php?title=Reeth *Contributors*: Achmelvic, Andre Engels, Baker46001, BrownHairedGirl, Cacolantern, CanOfWorms, Chris the speller, Cnyborg, Dpaajones, Euchiasmus, Farrtj, Harris000, Jh51681, Johnny64, Jpbowen, Kadsjdh, Keith D, Lupin, Mick Knapton, Mtaylor848, NiervaIreniSian, Nreckert, Pax:Vobiscum, Pit-yacker, Rehevkor, Richmondshire, Ringolad, Soupsportz, SteinbDJ, Veledan, Wereon, 25 anonymous edits

Telephone_booth *Source*: http://en.wikipedia.org/w/index.php?title=Telephone_booth *Contributors*: Ahoerstemeier, Ajuk, AlainV, Alex earlier account, Bashereyre, Biscuittin, Blankfaze, Bmprice2000, Bob f it, Boleslaw, BorgQueen, Branddobbe, Bryan Derksen, Calieber, Capitoline, Ccacsmss, Christopherlin, Chrome89, Colt .55, Comme le Lapin, CommonsDelinker, Cross porpoises, Dantadd, Darkieboy236, Ddxc, Deathawk, Deor, Deror avi, Dfmock, Doco, Dposse, DrFrench, Drat, Ed g2s, Edison, Edward, Ekko, Esperant, Eurosong, Eurou812trash, Everyking, Flaming.muskrats, G-Man, GangsterChic, Generica, George Ho, Gilliam, Greswik, Hahbie, Hbent, HisSpaceResearch, Hypocaustic, Immanuel goldstein, Infamouspineapple, Infrogmation, Jasper33, Jeffq, Jibbajabba, JidGom, Jim.henderson, JonHarder, Jren207, Juzeris, Jvlm.123, KKC, Kchishol1970, Keeves, Ken Gallager, Keycard, Kingturtle, LDHan, Leonrw, Lexor, Lironl, Lizzzs, Lohengrin1991, Lotje, Lyricmac, Manuel de Sousa, Martpol, McB, Menchi, Mhking, Mike Garcia, Milkandspoon, Minesweeper, Mintguy, Mmh, Mojo29, Moony1000, Mrwojo, Mytwocents, NYArtsnWords, Nakon, Nezzadar, Nidonocu, NoSeptember, Nommonomanac, Nopetro, Nv8200p, Nycstudent578, Odie5533, Ohnoitsjamie, Oknazevad, Patcat88, Pengyanan, Petersimison12, Phgao, Philwelch, PhreeStyler, Pixelface, Plasticup, Pmsyyz, Pokrajac, Poontanga4life, PunkdPanther, Quadell, Radiojon, Ranveig, Rhindle The Red, Rich Farmbrough, Richiekim, Rmhermen, Robertvan1, SQGibbon, Sewing, Sfouk, Shizhao, Sillyfolkboy, Sladen, Sprocket, Stefan Kühn, Stuartdavidcollier, Stuz, Tagishsimon, TharkunColl, Themfromspace, This Rainy Sunday, ThorstenS, TransUtopian, TreasuryTag, Ulf Abrahamsson, Unisouth, Veinor, VentrueCapital, VeryVerily, Viajero, WNivek, WarFox, Weydonian, Whitepaw, Woohookitty, Wykymania, Zir, Zxb, 141 anonymous edits

Tadcaster *Source*: http://en.wikipedia.org/w/index.php?title=Tadcaster *Contributors*: Acabashi, Acalamari, Achmelvic, Angusmclellan, Ann Stouter, Apollosm, Arakunem, ArglebargleIV, AvicAWB, Barticus88, Bgold4, Bigjim14, Bigkim14, Bill Thayer, Billlo, Bullshot, Can't sleep, clown will eat me, Cbuckley, Clanposse, Crimpshrine27, Dr Greg, DuncanHill, Edgieboy, Glacialfox, GreatWhiteNortherner, Greenshed, Grutness, Guoguo12, Harkey Lodger, Jevansen, Jhamez84, JuJube, Jza84, Keith D, Keith Edkins, LeighBCD, Lightmouse, Lozleader, Lupin, MRSC, Mais oui!, Mantavani, Marcallow, Materialscientist, Metricmike, Mhockey, Mike s, Morwen, Mtaylor848, Mushroom, N96, Neddyseagoon, Neewah, Nev, Nigelcoates, Noisy, Northumbrian, Nossidge, Orioane, Pit-yacker, PrestonH, Pschemp, RHaworth, Saga City, Saintsixtus, Sarumio, Scarian, Sloman, Snowolf, Stanismint, Stephenb, Taddylusc, Tellyaddict, TheParanoidOne, Theeaglehaslanded, Theroadislong, Tide rolls, Tingtongbingbongboo, Tonywalton, Tristanheaven, Vanished user 39948282, Vignaux, Vipinhari, WOSlinker, Waggers, Walgamanus, Warofdreams, Wereon, Whohe!, Yorkshire Phoenix, 91 anonymous edits

Reeth,_Fremington_and_Healaugh *Source*: http://en.wikipedia.org/w/index.php?title=Reeth%2C_Fremington_and_Healaugh *Contributors*: DINOMAN, Doprendek, DuncanHill, Keith D, Rich Farmbrough, Rodhullandemu, Waacstats

Image Sources, Licenses and Contributors

Image:153977 5721238c-Healaugh(CharlesRawding)Apr2006.jpg *Source*: http://en.wikipedia.org/w/index.php?title=File:153977_5721238c-Healaugh(CharlesRawding)Apr2006.jpg *License*: unknown *Contributors*: Charles Rawding

file:North Yorkshire UK location map.svg *Source*: http://en.wikipedia.org/w/index.php?title=File:North_Yorkshire_UK_location_map.svg *License*: unknown *Contributors*: User:Nilfanion

File:Red pog.svg *Source*: http://en.wikipedia.org/w/index.php?title=File:Red_pog.svg *License*: unknown *Contributors*: Anomie

File:Swaledale-web.jpg *Source*: http://en.wikipedia.org/w/index.php?title=File:Swaledale-web.jpg *License*: unknown *Contributors*: User:Cap

File:EastApplegarth.jpg *Source*: http://en.wikipedia.org/w/index.php?title=File:EastApplegarth.jpg *License*: unknown *Contributors*: James@hopgrove

Image: YorkshireDalesSign.jpg *Source*: http://en.wikipedia.org/w/index.php?title=File:YorkshireDalesSign.jpg *License*: unknown *Contributors*: Original uploader was LordHarris at en.wikipedia

Image: Yorkshire Dales National Park.png *Source*: http://en.wikipedia.org/w/index.php?title=File:Yorkshire_Dales_National_Park.png *License*: unknown *Contributors*: Original uploader was Keith Edkins at en.wikipedia. Later version(s) were uploaded by Jza84 at en.wikipedia.

File:Western Face of Thwaites scars.jpg *Source*: http://en.wikipedia.org/w/index.php?title=File:Western_Face_of_Thwaites_scars.jpg *License*: unknown *Contributors*: Childzy. Original uploader was Childzy at en.wikipedia

File:Gaping Gill.jpg *Source*: http://en.wikipedia.org/w/index.php?title=File:Gaping_Gill.jpg *License*: unknown *Contributors*: Original uploader was Mjobling at en.wikipedia

Image:Hawes house 01.JPG *Source*: http://en.wikipedia.org/w/index.php?title=File:Hawes_house_01.JPG *License*: unknown *Contributors*: User:Immanuel Giel

Image:Dry stone wall 20.JPG *Source*: http://en.wikipedia.org/w/index.php?title=File:Dry_stone_wall_20.JPG *License*: unknown *Contributors*: Immanuel Giel, Joadl, Ulrichstill

Image:Janet's Foss 2.jpg *Source*: http://en.wikipedia.org/w/index.php?title=File:Janet's_Foss_2.jpg *License*: unknown *Contributors*: User:Dbenbenn

Image:Ingleborough whole.JPG *Source*: http://en.wikipedia.org/w/index.php?title=File:Ingleborough_whole.JPG *License*: unknown *Contributors*: Original uploader was Childzy at en.wikipedia

File:Richmondshire arms.png *Source*: http://en.wikipedia.org/w/index.php?title=File:Richmondshire_arms.png *License*: unknown *Contributors*: Lozleader, Magul

File:NorthYorkshireRichmondshire.png *Source*: http://en.wikipedia.org/w/index.php?title=File:NorthYorkshireRichmondshire.png *License*: unknown *Contributors*: Keith D, Nichtbesserwisser

File:Honour of Richmond.png *Source*: http://en.wikipedia.org/w/index.php?title=File:Honour_of_Richmond.png *License*: unknown *Contributors*: User:Thomas Gun

File:North Yorkshire UK locator map 2010.svg *Source*: http://en.wikipedia.org/w/index.php?title=File:North_Yorkshire_UK_locator_map_2010.svg *License*: unknown *Contributors*: User:Nilfanion

File:Coat of arms of North Yorkshire County Council.png *Source*: http://en.wikipedia.org/w/index.php?title=File:Coat_of_arms_of_North_Yorkshire_County_Council.png *License*: unknown *Contributors*: User:Aups, User:Bruno Vallette, User:Gaeser, User:Jza84, User:Sodacan, User:Yorick, User:Zigeuner

Image:North Yorkshire Ceremonial Numbered.png *Source*: http://en.wikipedia.org/w/index.php?title=File:North_Yorkshire_Ceremonial_Numbered.png *License*: unknown *Contributors*: Michiel1972, Skinsmoke

Image:Bolton Abbey 7.jpg *Source*: http://en.wikipedia.org/w/index.php?title=File:Bolton_Abbey_7.jpg *License*: unknown *Contributors*: User:Dbenbenn

Image:Knaresborough Viaduct.jpg *Source*: http://en.wikipedia.org/w/index.php?title=File:Knaresborough_Viaduct.jpg *License*: unknown *Contributors*: User:TJBlackwell

Image:Dob Park Lodge.jpg *Source*: http://en.wikipedia.org/w/index.php?title=File:Dob_Park_Lodge.jpg *License*: unknown *Contributors*: User:TJBlackwell

File:North yorkshire moors railway map.gif *Source*: http://en.wikipedia.org/w/index.php?title=File:North_yorkshire_moors_railway_map.gif *License*: unknown *Contributors*: Original uploader was Maniac Pony (talk) at en.wikipedia. Later version(s) were uploaded by Achmelvic at en.wikipedia.

Image:Reeth from Fremington Edge.jpg *Source*: http://en.wikipedia.org/w/index.php?title=File:Reeth_from_Fremington_Edge.jpg *License*: unknown *Contributors*: Cnyborg, Rimshot, Thomas Gun

File:The Village of Reeth, North Yorkshire - geograph.org.uk - 226704.jpg *Source*: http://en.wikipedia.org/w/index.php?title=File:The_Village_of_Reeth,_North_Yorkshire_-_geograph.org.uk_-_226704.jpg *License*: unknown *Contributors*:

Image:Lightmatter phonebooths.jpg *Source*: http://en.wikipedia.org/w/index.php?title=File:Lightmatter_phonebooths.jpg *License*: unknown *Contributors*: Aaron Logan

Image:K6 Telephone Box and Edward VII Pillar Box Amberley.jpg *Source*: http://en.wikipedia.org/w/index.php?title=File:K6_Telephone_Box_and_Edward_VII_Pillar_Box_Amberley.jpg *License*: unknown *Contributors*: User:Unisouth

Image:ManUsingPhoneBox Footscray.JPG *Source*: http://en.wikipedia.org/w/index.php?title=File:ManUsingPhoneBox_Footscray.JPG *License*: unknown *Contributors*: Bukk, Leonrw

Image:Broken phone box.jpg *Source*: http://en.wikipedia.org/w/index.php?title=File:Broken_phone_box.jpg *License*: unknown *Contributors*: Edward

Image:Phone box prostitute calling cards 1.jpg *Source*: http://en.wikipedia.org/w/index.php?title=File:Phone_box_prostitute_calling_cards_1.jpg *License*: unknown *Contributors*: Edward, Fui in terra aliena, Infrogmation, Opponent, Oxyman, TwoWings, 1 anonymous edits

Image:Telephone booth 1 Sao Paulo Brasil.jpg *Source*: http://en.wikipedia.org/w/index.php?title=File:Telephone_booth_1_Sao_Paulo_Brasil.jpg *License*: unknown *Contributors*: user:Morio

Image:WW2 public phone.jpg *Source*: http://en.wikipedia.org/w/index.php?title=File:WW2_public_phone.jpg *License*: unknown *Contributors*: Original uploader was Keycard at en.wikipedia

Image:Tadcaster.jpg *Source*: http://en.wikipedia.org/w/index.php?title=File:Tadcaster.jpg *License*: unknown *Contributors*: Mtaylor848 (talk). Original uploader was Mtaylor848 at en.wikipedia

File:Tadcaster Social Club.jpg *Source*: http://en.wikipedia.org/w/index.php?title=File:Tadcaster_Social_Club.jpg *License*: unknown *Contributors*: User:Mtaylor848

File:Calcaria.jpg *Source*: http://en.wikipedia.org/w/index.php?title=File:Calcaria.jpg *License*: unknown *Contributors*: User:Mtaylor848

File:JS Brewery.jpg *Source*: http://en.wikipedia.org/w/index.php?title=File:JS_Brewery.jpg *License*: unknown *Contributors*: Mtaylor848 (talk). Original uploader was Mtaylor848 at en.wikipedia

File:Samuel Smiths Old Brewery.jpg *Source*: http://en.wikipedia.org/w/index.php?title=File:Samuel_Smiths_Old_Brewery.jpg *License*: unknown *Contributors*: User:Mtaylor848

File:Coors or bass brewery Tadcaster.jpg *Source*: http://en.wikipedia.org/w/index.php?title=File:Coors_or_bass_brewery_Tadcaster.jpg *License*: unknown *Contributors*: User:Mtaylor848

File:Tadcaster, Church of St Mary The Virgin - geograph.org.uk - 232926.jpg *Source*: http://en.wikipedia.org/w/index.php?title=File:Tadcaster,_Church_of_St_Mary_The_Virgin_-_geograph.org.uk_-_232926.jpg *License*: unknown *Contributors*: J.-H. Janßen

File:Viaduct - geograph.org.uk - 11601.jpg *Source*: http://en.wikipedia.org/w/index.php?title=File:Viaduct_-_geograph.org.uk_-_11601.jpg *License*: unknown *Contributors*: Ardfern, Oranjblud, Sf5xeplus

Printed by Books on Demand GmbH, Norderstedt / Germany